RECONCILING
GENESIS & SCIENCE
Unlocking the Theories of Creation

Fred Snowden

Library of Congress Control Number: 2021901194
ISBN-13:   Paperback: 978-1-64749-337-0
          ePub: 978-1-64749-338-7

Printed in the United States of America

GoToPublish LLC
1-888-337-1724
www.gotopublish.com
info@gotopublish.com

CONTENTS

PROLOGUE

I have spent much of my life and career mulling over the wonders of creation. All of us innately wonder about the existence of the Universe and of ourselves. Today, as with politics, Americans are sharply divided over the issues of evolution and creationism, but few ever stop to study these concepts thoroughly. My purpose is to put the most convincing theories, from all perspectives, one display and let the readers decide. Hopefully, what makes sense and what does not will become clear.

More importantly, my hopes are that proponents from both sides will relax and become less adversarial. It is understood that the questions surrounding the existence of the Universe are significant, but not to the point of making enemies of those who are only seeking truth about how it happened. Hopefully, an understanding of the opponents' motivations and attitudes may create some degree of empathy and compassion. It might even be possible for people with opposing views to become colleagues and friends.

Have you ever heard that modern man may have existed for over a million years?

Are you aware that there might be a vast difference between the orbital and atomic calendars?

Did you know that modern research suggests that our DNA is less like chimpanzees or Neanderthals and more like rats?

Can you believe that there is overwhelming evidence that the human brain is still evolving?

Have you considered that the mighty Zulus and the tiny Pygmies in Africa prove Darwin's theory of evolution?

This and much more awaits you; but first let us find out why we believe what we believe about evolution and creationism.

CHAPTER ONE

SEEKING ANSWERS FROM PHILOSOPHY

The modern controversy regarding Creation has become an out of control social phenomena. The conflict is driven by passion and fueled by prejudice. The battlefield is the school, bookstore, library, classroom, museum, church, mosque, temple and synagogue. The winners are those who profit from the war. The losers are school age students, sincere college students and the rest of us who just want factual information rather than defensive rhetoric, objectivity as opposed to blind passion, and transparency rather than a strategic cover-up. The shame is that there are sincere, intelligent people on both sides of the issue, but most are so caught up in defending their sacred ground that they will not even listen, much less consider each other's viewpoints. When the battle lines are finally drawn, the caissons line up, each with a different world view to protect and defend using the ammunition they hold.

On the extreme left are the secularists or objectivists who believe that the universe and all life is self-perpetuating through natural law and reject the idea of any intelligence or design behind creation. To them the Genesis narrative is primitive nonsense, comparable to Lord of the Rings or Jack and the Bean Stalk. Objectivism holds that reality is absolute regardless of one's fears, dreams, hopes or desires. It holds that ideas must be true, thus practical. Objectivism rejects that reality is determined by personal opinion or social convention, so that an individual's ideas or beliefs do not make reality what it is. There is no evidence or rational argument in support of the existence of a supernatural being who creates and/or controls reality, so objectivists must embrace other explanations for the existence of the Universe.

On the extreme right are the young earth creationists (YEC) who believe that God created a fully matured universe about 6,000 years ago, in six normal 24-hour days, with the light from the farthest stars already visible on earth, exactly as is described in the Genesis narrative. This position may also be called Literalism, which is the adherence to the explicit substance of an idea or expression and is most associated with Biblical interpretation. It equates to the dictionary definition of literalism, "adherence to the letter, word and literal sense, where literal means the strict sense of the word or words, not figurative or metaphorical."

To the right of the objectivists we find the Theistic Evolutionists, or compatibilists, who believe that the hand of a "Creator" initiated and guided the processes of evolution over eons of time. In their minds, the Big Bang and Darwinian Evolution are simply tools on the tool belt of the Creator that have produced what we observe today. For them, the Genesis narrative is but an understandable explanation of an overly complex operation, which the people in the day that the Bible was written could understand, given their limited understanding of science. Compatibilism offers a solution to the free will problem in that it holds that free will is compatible with determinism, in that free will is normally taken to be a necessary condition of moral responsibility.

Another step to the right takes us to Intelligent Design creationists (IDC), who believe that an intelligence created the universe, Earth, and each individually specific life form. These scientists are careful to stay away from words like God or Bible, so they appear to be guided by science, rather than religion. As is often pointed out by objectivists, there is a close relationship between intelligent design theory and traditional creationism or Old Earth Creationism (OEC). Most intelligent design theorists accept the old earth theory, and many accept some degree of common descent. The Discovery Institute, a privately supported think tank in Seattle, funds much of the work of the IDC movement. The driving force at the institute is University of Chicago educated young earth creationist and philosopher, Paul Nelson.

Between intelligent design creationists and the literalists, we find the first cousins of the young earth creationists, traditional creationists, or old earth creationists (OEC). They believe that the IDC's intelligence behind creation is the God of the Bible, and that the six "days" of the Genesis narrative are six eons of immeasurable time, not 24-hour days as the literalists insist. However, OECs would say they are the literalists and that their right-wing YEC cousins are hyper-literalists. Their belief in an "old earth" aligns them with objectivists and theistic evolutionists when it comes to geology, carbon dating, etc. Holding that reality has an inherently logical structure. Theists assert that truths exist, which the intellect can grasp directly without going to extremes. These rational principles in logic, ethics and even metaphysics are so fundamental that to deny them is to fall into contradiction. For traditional creationists, Darwinian evolution and YEC are two of those contradictions.

One might wonder, how there could be so many opinions about one issue, held by so many distinguished scholars of science, theology, and philosophy. The simple answer is that in the case of the creation, the first and most important step in the scientific method, 'observation', is missing. No one observed the actual creation. Without that foundational step, there is and will continue to be controversy regarding the process. You see, abstract ideas, opinions and personal perspective are a product of many mitigating factors, which begin in the individual mind at the molecular level. Thus, all human ideas are fueled by philosophy and world view, what one believes about life. Philosophy answers the BIG questions such as:

Am I a cosmic accident or a purposely designed individual?
Where did the human soul come from?
Why am I here?
What is my purpose?

Everyone has a philosophy or worldview, even if they don't recognize it or cannot express it in specific terms. Our philosophy or world view reveals itself as we live our lives, develop our values, and raise our children. The ancient Greeks understood philosophy better than any other culture and left to us volumes of material recognizing tendencies and ascribing descriptive names and characteristics. Very few of us today can tell you whether they are a Stoic or an Epicurean, but the trained eye can easily discern this from one's lifestyle and values.

World View plays a major role in choosing a creation belief. Objectivists, who are atheists or agnostics, have no choice but to reject the Genesis narrative, while most Christians, orthodox Jews, and Muslims have no choice but to accept it. A prime example is Dr. Francis Collins, the primary investigator and brilliant mind who unlocked the human genome. For much of his life, Dr. Collins was an avowed atheist and Darwinian evolutionist. Yet, later in life he became a Christian. Over time, as he studied the Bible, grew in faith, and his world view drastically changed. Faith in God and an understanding of the scriptures realigned what he had previously believed about creation. To accommodate both faith and evolution, he compatabilized that Theistic Evolution, evolution guided by the hand of the Creator not random chance, was the best explanation for what he has observed in nature and in the laboratory. With this particular creation belief, he was able to retain what he understood about evolution and was not forced to surrender his scientific objectivity or intellectual integrity.

Rational philosophy can only lead to four propositions regarding the Genesis narrative and these have been held through the ages in various forms:

- The Genesis narrative is a fable, there is no supernatural force in creation, everything that exists is a product of the force of nature at work (evolutionist, objectivist)

- The Genesis narrative is an allegorical, creation myth, which presents an understandable, unscientific explanation for the unlearned (theistic evolutionist, compatibilist)
- The Genesis narrative is superfluous since science demonstrates that there is intelligence behind creation and that evolution is contradictory and irrational (intelligent design creationists)
- The Biblical narrative is a literal, reasonable, scientific explanation for what we observe in nature (creationist, theist (OEC), literalist (YEC).

Before moving on, it is important to distinguish between the terms, belief, and knowledge. Belief may be abstract and unsupported by fact or reason. It is mostly associate with religion and is respected, even protected, but not held as truth. Webster defines belief as "a feeling of being sure that something exists or is true, right or valuable". Knowledge is on a different level than belief; it is held as science that is supported by a precise set of facts and observations. All the positions previously listed would say that their understanding of creation is based in supportable fact and is thus is knowledge, not simply belief. For this reason, each must be held accountable to a higher standard. The intent of this book is to scrutinize the facts to which each group clings in an effort to sort out fact from fiction, knowledge from belief and truth from myth.

Now let us examine these positions, beginning with the domain of objectivism, which includes all those who never consider the possibility of a creator. For them, there is a naturalistic explanation for creation, so they must develop a scientific model that is completely independent of the supernatural. Objectivists are typified as unbridled liberals, but Ayn Rand, a social and economic conservative defined her philosophy as objectivism and called it "a philosophy for living on earth". Objectivism is an integrated system of thought that defines the abstract principles by which a man must think and act if he is to live the life proper to mankind. Like Homer, Ayn Rand first portrayed her philosophy in the form of the heroes of her best-selling novels, *The Fountainhead* (1943) and *Atlas Shrugged* (1957). She later expressed her philosophy in nonfiction form. She was once asked if she could present the essence of objectivism in a nutshell. Her answer was:

Metaphysics:	Reality
Epistemology:	Reason
Ethics:	Self-interest
Politics:	Capitalism

She then translated those terms into familiar language:
"Nature, to be commanded, must be obeyed."
"You can't eat your cake and have it, too."
"Man is an end in himself."
"Give me liberty or give me death."

For the objectivist, reality is evidenced by the external world, independent of man's consciousness, independent of any observer's knowledge, beliefs, feelings, desires, or fears. This means that 2+2 always equals 4, facts are facts, things are what they are, and the task of man's consciousness is to perceive reality, not to create or invent it. Thus, objectivism rejects any belief in the supernatural and any claim that individuals or groups can create their own reality. Objectivism also rejects mysticism, the idea that any acceptance of faith or feeling as a means of knowledge, and skepticism, the claim that certainty or knowledge is impossible. Man is a rational being and reason, as mankinds only means of knowledge, is his basic means of survival. However, the exercise of reason depends on each individual's choice. This one choice controls all the other choices you make and determines your life and character. Objectivism also rejects any form of determinism, the belief that man is a victim of forces beyond his control, such as God, fate, upbringing, genes, or economic conditions. Reason is man's only proper judge of values and his only proper guide to action. The proper standard of ethics is that which is required by man's nature for his survival as a rational being, rather than as a mindless brute.

Compatibilists, expressed in the previously mentioned Francis Collins, rationally examine the facts, and see an intelligent, guiding hand or supernatural force in creation. Thus, they are convinced that true science supports such a position even if the specifics of that force cannot be defined. Hume explained that the Compatibilists' free will should not be understood as some kind of ability to have actually chosen differently in an identical situation. The Compatibilist believes that a person always makes the only decision possible. Any talk of alternatives is strictly hypothetical. Faced with the standard argument against free will, many compatibilists choose determinism so that their actions are adequately determined by their philosophy, reasons, motives, culture, etc. Objectivists, also described as Hard Determinists, accuse Compatibilists, also described as Soft Determinists, of being motivated by a lack of a coherent, consonant moral belief system.

William James accused them of creating a "quagmire of evasion" by stealing the name of freedom to mask their underlying determinism. Immanuel Kant called it a "wretched subterfuge" and "word jugglery". Kant's argument turns on the view that, while all empirical phenomena must result from determining causes, human thought introduces something seemingly not found elsewhere in nature, the ability to conceive of the world in terms of how it *ought* to be, or how it might otherwise be. Because of its capacity to distinguish *is* from *ought*, reasoning can spontaneously originate new events without being itself determined by what already exists. Kant proposes that to take the compatibilist view is to deny the distinctly subjective capacity to re-think an intended course of action in terms of what *ought* to happen.

Literalists or theists are certain that specific supernatural revelation, such as the Tora, Quran or Bible, provides the only trustworthy explanation for creation. Most Evangelical Christians, Orthodox Jews, Muslims, and some Hindu sects are Creationists, in this sense. The Creationist believes in order, design, and therefore a designer. For them, this view has been constantly borne out by the new discoveries of the last century, of which the genome is just one. This intricate blueprint, that is unique to each individual and indeed is the makeup of all living organisms, could not have developed by chance or accident. Therefore, a creationist is one who trusts in God rather than the proverbial primeval soup as the as creator of all life on earth.

Creationists take comfort in the accomplishment of miracles by a Higher Power. Since they believe that God is omnipotent, evolution in any form cannot be valid. It is but an unproven theory that is subject to constant revision, rendering it almost laughable. They are completely comfortable with facing the obvious conflicts held to be 'modern reason' and are willing to be described as 'flat-earthers' by those who reject the idea of a supernatural creation because they understand how it is motivated. Most creationists are also castrophists and depend on other supernatural occurrences, like those mentioned in the Bible, to support their claims, especially the global deluge. They take confidence that such an event, which is also mentioned in extra-Biblical literature from almost every ancient culture in the world, really happened.

Finally, many creationists see a strong link between evolution and hotly contested current moral issues: abortion on demand, animal rights and homosexuality as a genetic predisposition. Since many literal six-day creationists are also dispensationalists, that God is working through several distinct periods of time to complete the history of mankind, the creation model best fits their view of eschatology. C.S. Lewis summed up the philosophy for Theism in this manner:

"One absolutely central inconsistency ruins the popular scientific philosophy. The whole picture professes to depend on inferences from observed facts. Unless inference is valid, the whole picture disappears... unless Reason is an absolute, all is in ruins. Yet those who ask me to believe this world picture also ask me to believe that Reason is simply the unforeseen and unintended by-product of mindless matter at one stage of its endless and aimless becoming. Here is flat contradiction. They ask me at the same moment to accept a conclusion and to discredit the only testimony on which that conclusion can be based." C.S. Lewis, *"Is Theology Poetry?"*

This argument against naturalism and materialism holds that:
1. No belief is rationally inferred if it can be fully explained in terms of non-rational causes.
2. If naturalism is true, then all beliefs can be fully explained in terms of non-rational causes. In other words, they can be explained by factors in nature, such as the four governing forces.
3. Therefore, if naturalism is true, then no belief is rationally inferred.
4. If any thesis entails the conclusion that no belief is rationally inferred, then it should be rejected and its denial accepted.

Conclusion: Therefore, naturalism should be rejected, and its denial accepted.

The argument for the existence of a Creator holds:

- A being requires a rational process to assess the truth or falsehood of a claim (hereinafter, to be convinced by argument).
- Therefore, if humans are able to be convinced by argument, their reasoning processes must have a rational source.
- Considering element two above, if humans can be convinced by argument, their reasoning processes must have a non-physical as well as a rational source.
- Rationality cannot arise out of non-rationality. That is, no arrangement of non-rational materials creates a rational thing.
- No being that begins to exist can be rational except through reliance, ultimately, on a rational being that did not begin to exist. That is, rationality does not arise spontaneously from out of nothing but only from another rationality.
- All humans began to exist at some point in time.
- Therefore, if humans are able to be convinced by argument, there must be a necessary and rational being on which their rationality ultimately relies.
- Conclusion: This being we call God.

The Islamic scholar, Hamza Andreas Tzortzis, put it this way:

"No question is more sublime than why there is a universe: why there is anything rather than nothing. When we reflect upon our own existence we will come to the realization, that at some point in time, we began to exist. Since we were once non-existent and are now in existence, it follows that we must have had a beginning. In light of this, the Qur'an raises some profound questions".

"Were we created by nothing? Did we create ourselves? Or did we create the universe? Or were they created by nothing? Or were they the creators (of themselves)? Or did they create heavens and earth? Rather, they are not certain." Quran 52:35-36

Objectivists would say that Creationists, through a process known as cognitive dissonance, the state of holding two or more conflicting ideas, beliefs, values and/or emotional reactions, accommodate reality. That they achieve a philosophical balance or psychological awareness of uncomfortable tension caused by holding two conflicting thoughts or beliefs in one mind at the same time. Since they tend to seek consistency in their beliefs and perceptions, something must change in order to eliminate or reduce the dissonance. The theory of cognitive dissonance in social psychology proposes that people have a motivational drive to reduce dissonance by altering existing cognitions, adding new ones to create a consistent belief system, or alternatively by reducing the importance of any one of the dissonant elements.

Creationists, on the other hand, might say that it is the objectivists who allow cognitive dissonance to comfort them in their 'delusion'. A delusion because they are anthropologically aware of the existence of a Creator but reject him in order to escape a life lived in submission to such an invisible higher power. That in the absence of genuine proof of life arising from non-living matter, of undisputable transitional forms in the fossil record and that in the face of the overwhelming mathematical improbability that man could evolve from lower forms, they cling to their 'faith' in nature and blindly refuse to consider that there could be intelligence behind creation.

If we agree that one's world view or philosophy is the driving factor in what we believe about creation, we can also agree that world view creates more questions than answers. Philosophy is totally subjective, so it seems as if our creation beliefs are simply a reinforcement of our philosophical bias. Thus the contempt, vehemence and downright subterfuge that take place in the schools, media and laboratories are fueled more by philosophy than pure science, so the combatants can never surrender. The validated facts, that are held in common by all sides, are interpreted with particular philosophical bias. If one does not have good evidence, we fake it until we do. It is always "too soon or too late to tell". If we fail, we cover it up and dig it up again in at the same spot, hoping for a new twist. Intelligent scientists, philosophers and theologians are thus reduced to dogs chasing their own tails.

So, what are the seekers of truth to do? We must make a sincere attempt to sort out fact from fiction, science from theory and sound theology from blind faith. There is room for debate, there is room for compromise but there is no longer room for the 'soapbox'. It is time for a truce in which all combatants stop bickering and start listening to one another in an effort to generate a genuine, empathetic understanding of each other's point of view. Agreement is not a realistic goal, but a willingness to admit that a person's reality is not flawed simply because he/she does not agree about an event that was never observed. To do this, a rational, unvarnished understanding of the scientific method is critical to exploring the subjects and issues that will be discussed hereafter. The elements of this system are:

OBSERVATION – GATHERING DATA
- Looking at Nature
- Quantifying & organizing data
- Researching what others have observed

HYPOTHESIS/THEORY–FORMULATING A REASONABLE EXPLANATION
- Must be Testable (practical)
- Must be Reproducible over a period of time

EXPERIMENTATION – AN ARTIFICIAL, CONTROLLED INVESTIGATION
- Must have a Constant – that which does not change
- Must have a Variable – that which changes

ONLY AFTER A HYPOTHESIS IS PROVEN THROUGH MANY EXPERIMENTS & HAS THE ENDORSEMENT OF OTHER SCIENTISTS DOES IT BECOME A THEORY

WHEN A THEORY ENDURES THE TEST OF TIME, IT BECOMES SCIENTIFIC LAW

THERE HAVE BEEN OCCASIONS WHEN ACCEPTED SCIENTIFIC LAWS HAVE NEEDED TO BE REDEFINED, AS WITH BODE'S LAW

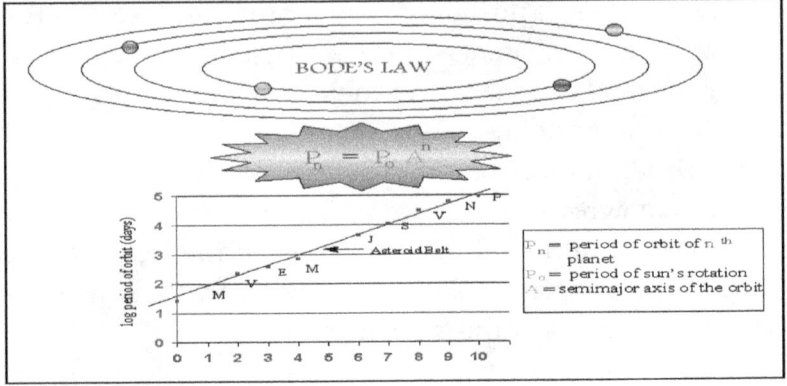

Bode was an accomplished mathematician, astronomer and physicist who was hindered by the limitations of the telescopes available in his day. He observed certain patterns in the orbits of the planets and quantified these observations mathematically. In applying his law, he predicted that there must be a hidden planet between Mars and Jupiter. His hypothesis made reasonable sense and was investigated by other scientists, which elevated his prediction to a theory. When the Asteroid Belt was discovered, the scientists of that day thought that they had proven Bode's theory and it became scientific law. With improvement to the telescope, however, Saturn was discovered and later Neptune, which better explained Bode's predictions and his prediction of the hidden planet was dismissed. This does not negate his genius, but it does make a good point; even established scientific law may be open to modification as better instruments and new discoveries give rise to new evidence. This brings us to the limitations of science:

- Science is limited to proving predictions/theories related to matter and energy (the physical universe) that exist in the present.
- Things that are spiritual, supernatural, moral or mystical cannot be quantified – these belong to the philosophy of Metaphysics
- Things that occurred in the past and that might occur in the future cannot be observed or tested, so for now, theories or predictions dependent upon them must remain unproven
- Artifacts, archeological digs and peering into deep space give us clues to the past, but certain assumptions may still be required,

which are not science, even if those assumptions are held for long periods of time

- Although philosophical or world view bias should not affect or scientific objectivity, it often does:

We often accept theory as fact

We often accept long held assumptions as fact

We often accept what cannot possibly be proven as fact

CHAPTER TWO

SEEKING ANSWERS FROM THE BEGINNING

The first words in the Bible are: "In the beginning", yet, both creationists and evolutionists have long asked, when was that beginning? At what point did the envelope which we call time and space come into being? When did matter, energy and the four governing forces make their appearance? Bishop Ussher, a theologian, fixed the creation date as 4004 BC by using generational records from scripture, but most physicists and astronomers today argue that the universe is between 13.7-13.8 billion years old.

In 1927, Georges Lemaître first noted that an expanding universe could be traced back in time to an originating single point. For much of the rest of the 20th century scientific community was divided between supporters of the Big Bang and the rival steady-state model, but a wide range of empirical evidence strongly favored the Big Bang, which is now universally accepted. Edwin Hubble concluded from analysis of galactic redshifts in 1929 that galaxies are slowly drifting apart; this is important observational evidence for an expanding universe.

Astronomers and physicist have become very dogmatic about the accuracy of their timing of the beginning of the universe since the discovery of the 'red shift' in space, which, as far as most are concerned, proved the theory of an expanding universe. Starting with Hawking and based on recent discoveries by the Hubble Telescope, scientists claim to have been able to reverse the expansion of the universe from the present to the moment at which from nothing, the universe came into existence. They say that they now have the tools and ability to, in a manner of speaking, rewind the creation event from the present to the beginning. It is somewhat like rewinding a movie on a DVD. When we push the rewind button, we are able go from the end to the beginning and observe the images in the order things actually happened. Scientists contend that what has been discovered through this process is equivalent to observation itself; others disagree.

This is not the only evidence that modern objectivists rely upon to estimate the age of the universe. For generations astronomers have been observing and gaining understanding the life cycle of stars. The evidence from observation is that stars have a life cycle very similar to humans; they are born, they live through distinguishable stages of life, they flame-out and die, leaving behind the shell of their existence. It is through this process that objectivists believe the elements, those found in our Earth and other planets and asteroids, are created. If, in fact, the elements that constitute all matter are created in the stars, it is easy to understand that a literal 24-hour day interpretation of the Genesis narrative seems totally unreasonable.

The 'star story' is quite amazing; how that from the simplest element, hydrogen, all of the 98 naturally occurring elements are assembled. An understanding of the life cycle of Stars helps us understand that the elements, up to the mass of iron, are made in stars and how the elements with greater mass than iron, are formed in the Supernova stage, which also distributes these elements throughout the universe. The remaining, manmade elements are assembled from the naturally occurring elements, which are commonly arranged in this illustrative chart.

Group →	1	2	3	4	5	6	7	8	9	10	11	12	13	14	15	16	17	18
↓ Period																		
1	1 H																	2 He
2	3 Li	4 Be											5 B	6 C	7 N	8 O	9 F	10 Ne
3	11 Na	12 Mg											13 Al	14 Si	15 P	16 S	17 Cl	18 Ar
4	19 K	20 Ca	21 Sc	22 Ti	23 V	24 Cr	25 Mn	26 Fe	27 Co	28 Ni	29 Cu	30 Zn	31 Ga	32 Ge	33 As	34 Se	35 Br	36 Kr
5	37 Rb	38 Sr	39 Y	40 Zr	41 Nb	42 Mo	43 Tc	44 Ru	45 Rh	46 Pd	47 Ag	48 Cd	49 In	50 Sn	51 Sb	52 Te	53 I	54 Xe
6	55 Cs	56 Ba		72 Hf	73 Ta	74 W	75 Re	76 Os	77 Ir	78 Pt	79 Au	80 Hg	81 Tl	82 Pb	83 Bi	84 Po	85 At	86 Rn
7	87 Fr	88 Ra		104 Rf	105 Db	106 Sg	107 Bh	108 Hs	109 Mt	110 Ds	111 Rg	112 Cn	113 Uut	114 Fl	115 Uup	116 Lv	117 Uus	118 Uuo

Lanthanides	57 La	58 Ce	59 Pr	60 Nd	61 Pm	62 Sm	63 Eu	64 Gd	65 Tb	66 Dy	67 Ho	68 Er	69 Tm	70 Yb	71 Lu
Actinides	89 Ac	90 Th	91 Pa	92 U	93 Np	94 Pu	95 Am	96 Cm	97 Bk	98 Cf	99 Es	100 Fm	101 Md	102 No	103 Lr

For the sake of simplicity, we might compare these elements to a Lego set, which is comprised of connecting pieces of different sizes, shapes, and colors. From these Legos, almost anything imaginable can be built.

According to physicists, spectrographic evidence verifies that stars and Supernova are forming elements right now, so it becomes evident that the building blocks of the universe or 'God's Lego set' are produced in the stars. It seems that an abundant amount of Hydrogen (H) was created at the onset of the Big Bang. We must also assume that the four governing forces came into existence at that same moment. As the hydrogen spread throughout the universe, the force of gravity caused the hydrogen atoms to collect in massive Nebula clouds. As a Nebula develops, rotation follows, and the hydrogen molecules get closer and closer together. The friction between molecules generates heat. Over time the temperature reaches critical mass, and a new star is born.

In stars that generate temperatures greater than 15,000,000 degrees F, a nuclear chain reaction continues, and the process of nuclear fusion begins to fuse 2 hydrogen atoms into one Helium (He) atom. In this process a tremendous amount of energy is released. This is the process that was harnessed to produce fusion nuclear weapons and nuclear power plants. In high mass stars reaching the Red Giant phase, where enough heat is available, the process continues as carbon, neon, oxygen, silicon, and iron are formed.

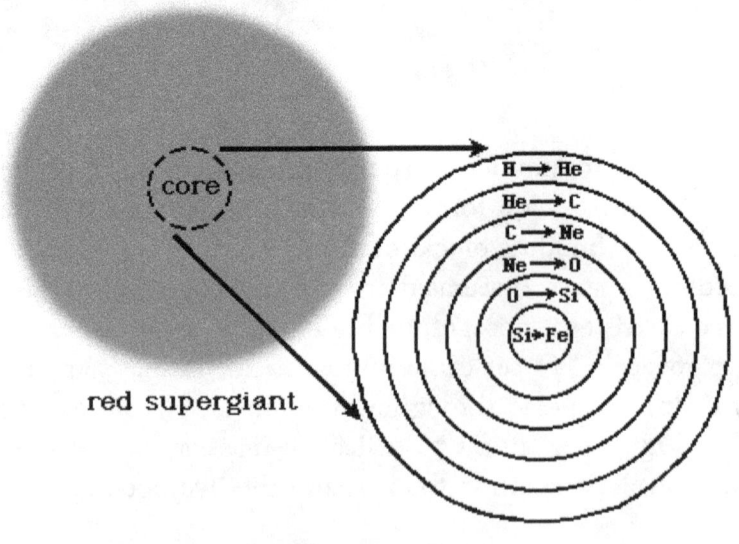

Once again gravity enters the picture and the iron concentrates in the core of the star, causing fusion to slow and eventually cease. A battle between the cold contracting mass of iron and the superheated expanding gasses ensues. When temperatures approach 10 billion degrees F, the star collapses resulting in an incredible explosion called Supernova. In the heat of that explosion physicists say the elements heavier than iron are formed.

We might call this the physicist's version of material evolution, thus giving rise to the "old earth" theory, which is compatible with Darwinian evolution. Certainly, their evidence seems credible and verifiable and lends itself to a reasoned explanation of how the universe formed. So, even though no one could observe the formation of the elements "in the beginning", we can observe that process in real time, right now; relegating it to science, not speculation. An understanding of nuclear fusion forms a firm scientific foundation for understanding how the stars are, in fact, element factories, However, it leads creationists and evolutionists back to the instant before time began to the unanswered question. From where did time, space, matter, energy and four fundamental governing forces come? Objectivists say it was inevitability, IDC says it was intelligence and theists say it was the Creator.

The Big Bang changed everything for creationists! The universe had not always existed, as scientists previously assumed, it had a birth. From nothingness, space, time, energy, matter and the four governing forces instantaneously came into being as the Genesis narrative declares. The dynamics of the Big Bang also explain two difficult to understand creation scriptures, which are unlocked by Gary Driver in his article published in Ancient Origins, February 28, 2020.

1. *"Through faith we understand that the worlds were framed by the word of God, so that things which are seen were not made of things which do appear." Hebrews 11:3*

   By knowledge obtained through observation, we know the governing forces undergirding creation were established within the first instant of creation along with space, time, energy, and matter. "Let there be..." not only refers to His fiat (command), but also to the information necessary to accomplish the command. Now that we better understand the creation events, Romans 1:20 can be better understood. This verse was mostly known by the general interpretation of the invisible attributes, eternal power and divine nature of God manifested in creation at large. However, upon an in-depth investigation of the heavens, an additional revelation emerges.

2. *"For the **invisible things** of him from the creation of the world **are clearly seen**, being understood by the things that are made, even his eternal power and Godhead; so that they are without excuse."*

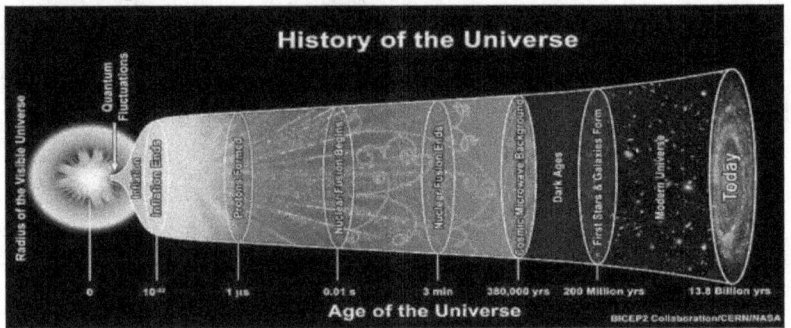

*The context of both these scriptures is the creation of the universe, which we can see and observe. The reference to invisible things (plural) was matter and energy, not yet in the visible spectrum. These invisible things are proclaimed to be invisible at the very start of time. So, how can the physical things that are referenced be both **invisible and clearly seen**? At birth, the universe was in the extreme high energy gamma ray part of the electromagnetic spectrum, making it invisible to humans. Because of an extremely high temperature, estimated to be 1032 Kelvin at creation, the constituents of mass could not exist as a unit (atoms) resulting in an opaque, glowing matter plasma; "Let there be light". The matter in creation could not endure as physical entities (atoms) since at such a high temperature, as soon as it came into being. We might say they were "virtual particles". Due to the apparent absence of physical particles, it seemed like there was nothing, but in reality, everything needed to build the universe was present (1st law of thermodynamics).*

As temperatures cooled, atoms could bind (cosmic recombination) and space became transparent, allowing photons to be emitted and detected as visible cosmic microwave background radiation. The universe was expanding (the red shift) expanded and cooled, the particles came into being and remained; eventually, the stars and galaxies formed. The scripture declares the original creation condition changed from invisible to now clearly seen (Greek - to see with the eyes). The appearance of photon light is the - "let there be light" in Genesis 1:3. This is not in conflict with the creation of our Sun being formed from existing material one day (Yom) four (Genesis 1:14). Thus, creation was initially invisible (energy) and is now clearly seen (mass). The expansion of space/time in tensed time allowed the invisible energy to cool and transform into visible matter, affirming E=MC2. This is the duality of invisible energy and the clearly seen mass spoken of in the scripture.

NASA

So, even though no one could observe the formation of the elements "in the beginning", we can observe that process in real time, right now; relegating it to science, not speculation. An understanding of nuclear fusion forms a firm scientific foundation for understanding how the stars are, in fact, element factories, However, it leads creationists and evolutionists back to the instant before time began to the unanswered question. From where did time, space, matter, energy and four fundamental governing forces come? Objectivists say it was inevitability, IDC says it was intelligence and theists say it was the Creator.

Once again, it is not the science, but the interpretation of the event that makes Professor Hawking's 'Theory of Everything' become more viewpoint or philosophy than science. Let's be clear! Evolutionists and Creationists now agree that the Universe has not always existed and that in an instant, space, time, matter, energy and the four governing forces came into existence. It seems that Dr. Stephen Hawking became an unwilling participant in bringing creationists and evolutionists together in his refinement of the Singularity Principle.

Traditional creationists do not get hung up on specific timing issues since they either do not hold to a literal six-day creation period as they have created various theological 'gap theories' to span enough time to be able to accept that the universe could be billions of years old. These old earth creationists approach the creation accounts of Genesis in a number of different ways. The framework interpretation or framework hypothesis notes that there is a pattern or "framework" present in the Genesis account and that, because of this, the account may not have been intended as a strict chronological record of creation. Instead, the creative events may be presented in a topical order. This view is broad enough that proponents of other old earth views, including many Day-Age creationists, have no problem with many of the key points put forward by the hypothesis.

Summary of the Genesis 6-day creation account, showing the pattern according to the framework hypothesis.
From - http://www.thefullwiki.org/Answers_in_Creation

Days of creation	Days of creation
Day 1: Light; day and night	**Day 4:** Sun, moon and stars
Day 2: Sea and Heavens	**Day 5:** Sea creatures; birds
Day 3: Land and vegetation	**Day 6:** Land creatures; man

Day-age creationism is an effort to reconcile the literal Genesis account of creation with modern scientific theories on the age of the universe, the Earth, life, and humans. It holds that the six days referred to in the Genesis account of creation are not ordinary 24-hour days, but rather are much longer periods of time.

Gap theories are quite another thing. Gap theologians create vast holes in the history of the universe to span immense periods of time. The most popular gap theory is that there is a gap of indeterminate time between Genesis 1:1 and 1:2. One of the most developed theories proposed by the Churches of God Worldwide goes like this:

That there was a pre-Adamic creation seems undeniable. The assertion, however, that this is contrary to the biblical narration does not stand up to scrutiny. One of the most fundamental errors of the theologians was that of Genesis chapter 1, verses 1-2. The term, "was waste and void" does not denote a condition of creation, but rather a condition of destruction. The earth became tohuw and bohuw, or waste and void; it was not created that way (cf. Isaiah 45:18). Simply put, the earth has existed for many millions of years. The Genesis story is a re-creation story, not the total creation story. The gap between Genesis 1:1 and Genesis 1:2 is untold millions of years.

This is but one of hundreds of theological explanations for gaps that allow traditional creationists to accept an old earth, but as was noted before; young earth creationists believe that the Earth could not possibly be more than 6,000- 10,000 years old. Their model is also based in science, not faith. They have been in the laboratory and discovered their own body of evidence, as is illustrated by this Creation Research Institute presentation developed by Russell Humphries, Ph.D.:

1. Galaxies wind themselves up too fast.

The stars of our own galaxy, the Milky Way, rotate about the galactic center with different speeds, the inner ones rotating faster than the outer ones. The observed rotation speeds are so fast that if our galaxy were more than a **few hundred million years old**, it would be a featureless disc of stars instead of its present spiral shape.[1] Yet our galaxy is supposed to be at least 10 billion years old. Evolutionists call this «the winding-up dilemma,» which they have known about for fifty years. They have devised many theories to try to explain it, each one failing after a brief period of popularity. The same «winding-up» dilemma also applies to other galaxies. For the last few decades the favored attempt to resolve the puzzle has been a complex theory called «density waves.»[1] The theory has conceptual problems, has to be arbitrarily and very finely tuned, and has been called into serious question by the Hubble Space Telescope›s discovery of very detailed spiral structure in the central hub of the «Whirlpool» galaxy, M51.

2. Too few supernova remnants.

According to astronomical observations, galaxies like our own experience about one supernova (a violently-exploding star) every 25 years. The gas and dust remnants from such explosions (like the Crab Nebula) expand outward rapidly and should remain visible for over a *million years.* Yet the nearby parts of our galaxy in which we could observe

Crab Nebula
Photo: Courtesy of NASA

such gas and dust shells contain only about 200 supernova remnants. That number is consistent with only about **7,000 years**' worth of supernovas.

3. Comets disintegrate too quickly.

According to evolutionary theory, comets are supposed to be the same age as the solar system, about five billion years. Yet each time a comet orbits close to the sun; it loses so much of its material that it could not survive much longer than about 100,000 years. Many comets have typical ages of less than **10,000** years.4 Evolutionists explain this discrepancy by assuming that (a) comets come from an unobserved spherical "Oort cloud" well beyond the orbit of Pluto, (b) improbable gravitational interactions with infrequently passing stars often knock comets into the solar system, and (c) other improbable interactions with planets slow down the incoming comets often enough to account for the hundreds of comets observed.5 So far, none of these assumptions has been substantiated either by observations or realistic calculations. Lately, there has been much talk of the «Kuiper Belt,» a disc of supposed comet sources lying in the plane of the solar system just outside the orbit of Pluto. Some asteroid-sized bodies of ice exist in that location, but they do not solve the evolutionists› problem, since according to evolutionary theory, the Kuiper Belt would quickly become exhausted if there were no Oort cloud to supply it.

4. Not enough mud on the sea floor.

Each year, water and winds erode about 20 billion tons of dirt and rock from the continents and deposit it in the ocean.[6] This material accumulates as loose sediment on the hard basaltic (lava-formed) rock of the ocean floor. The average depth of all the sediment in the whole ocean is less than 400 meters.[7] The main way known to remove the sediment from the ocean floor is by plate tectonic subduction. That is, sea floor slides slowly (a few cm/year) beneath the continents, taking some sediment with it. According to secular scientific literature, that process presently removes only 1 billion tons per year.[7] As far as anyone knows, the other 19 billion tons per year simply accumulate. At that rate, erosion would deposit the present mass of sediment in less than **12 million years**. Yet according to evolutionary theory, erosion and plate subduction have been going on as long as the oceans have existed, an alleged three billion years. If that were so, the rates above imply that the oceans would be massively choked with sediment dozens of kilometers deep. An alternative (creationist) explanation is that erosion from the waters of the Genesis flood running off the continents deposited the present amount of sediment within a short time, about 5,000 years ago.

Rivers and dust storms dump mud into the sea much faster than plate tectonic sub-duction can remove it.

5. Not enough sodium in the sea.

Every year, rivers[8] and other sources[9] dump over 450 million tons of sodium into the ocean. Only 27% of this sodium manages to get back out of the sea each year.[9,10] As far as anyone knows, the remainder simply accumulates in the ocean. If the sea had no sodium to start with, it would have accumulated its present amount in less than 42 million years at today's input and output rates.[10] This is much less than the evolutionary age of the ocean, three billion years. The usual reply to this discrepancy is that past sodium inputs must have been less and outputs greater. However, calculations that are as generous as possible to evolutionary scenarios still give a maximum age of only **62 million** years.[10] Calculations11 for many other seawater elements give much younger ages for the ocean.

6. The earth's magnetic field is decaying too fast.

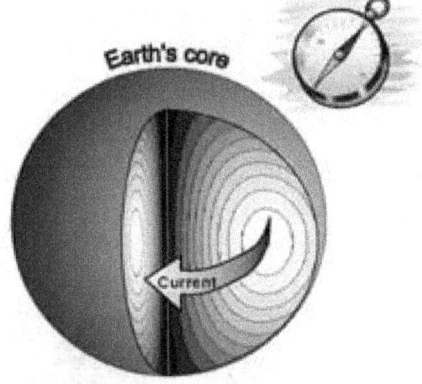

The total energy stored in the earth's magnetic field ("dipole" and "non-dipole") is decreasing with a half-life of 1,465 (± 165) years.[12] Evolutionary theories explaining this rapid decrease, as well as how the earth could have maintained its magnetic field for *billions of years* are very complex and inadequate. A much better creationist theory exists. It is straightforward, based on sound physics, and explains many features of the field: its creation, rapid reversals during the Genesis flood, surface intensity decreases

Electrical resistance in the earth's core wears down the electrical current which produces the earth's magnetic field. That causes the field to lose energy rapidly.

and increases until the time of Christ, and a steady decay since then.[13] This theory matches paleomagnetic, historic, and present data, most startlingly with evidence for rapid changes.[14] The main result is that the field's total energy (not surface intensity) has always decayed at least as fast as now. At that rate, the field could not be more than **20,000 years** old.[15]

7. Many strata are too tightly bent.

In many mountainous areas, strata thousands of feet thick are bent and folded into hairpin shapes. The conventional geologic time scale says these formations were deeply buried and solidified for *hundreds of millions* of years before they were bent. Yet the folding occurred without cracking, with radii so small that the entire formation had to be still wet and unsolidified when the bending occurred. This implies that the folding occurred less than thousands of years after deposition.[16]

8. Biological material decays too fast.

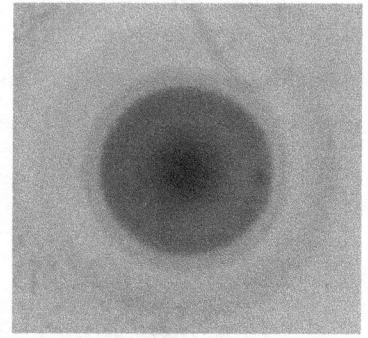

Natural radioactivity, mutations, and decay degrade DNA and other biological material rapidly. Measurements of the mutation rate of mitochondrial DNA recently forced researchers to revise the age of "mitochondrial Eve" from a theorized 200,000 years down to possibly as low as 6,000 years.[17] DNA experts insist that DNA cannot exist in natural environments longer than 10,000 years, yet intact strands of DNA appear to have been recovered from fossils allegedly much older: Neanderthal bones, insects in amber, and even from dinosaur fossils.[18] Bacteria allegedly 250 million years old apparently have been revived with no DNA damage.[19] Soft tissue and blood cells from a dinosaur have astonished experts.[20]

9. Fossil radioactivity shortens geologic "ages" to a few years.

Radio halos are rings of color formed around microscopic bits of radioactive minerals in rock crystals. They are fossil evidence of radioactive decay.[21] «Squashed» Polonium-210 radio halos indicate that Jurassic, Triassic, and Eocene formations in the Colorado plateau were deposited **within months** of one another, not *hundreds of millions of years* apart as required by the

Radio Halo, Photo: Courtesy of Mark Armitage

conventional time scale.[22]"Orphan" Polonium-218 radio halos, having no evidence of their mother elements, imply **accelerated nuclear decay** and very rapid formation of associated minerals.[23,24]

10. Too much helium in minerals.

Uranium and thorium generate helium atoms as they decay to lead. A study published in the Journal of Geophysical Research showed that such helium produced in zircon crystals in deep, hot Precambrian granitic rock has not had time to escape.[25] Though the rocks contain 1.5 billion years' worth of nuclear decay products, newly-measured rates of helium loss from zircon show that the helium has been leaking for only **6,000** (± 2000) years.[26] This is not only evidence for the youth of the earth, but also for episodes of greatly accelerated decay rates of long half-life nuclei within thousands of years ago, compressing radioisotope timescales enormously.

11. Too much carbon 14 in deep geologic strata.

With their short 5,700-year half-life, no carbon 14 atoms should exist in any carbon older than 250,000 years. Yet it has proven impossible to find any natural source of carbon below Pleistocene (Ice Age) strata that does not contain significant amounts of carbon 14, even though such strata are supposed to be *millions or billions of years* old. Conventional carbon 14 laboratories have been aware of this anomaly since the early 1980s, have striven to eliminate it, and are unable to account for it. Lately the world's best such laboratory which has learned during two decades of low-C14 measurements how not to contaminate specimens externally, under contract to creationists, confirmed such observations for coal samples and even for a dozen diamonds, which cannot be contaminated in situ with recent carbon.[27] These constitute very strong evidence that the earth is only **thousands**, not billions, of years old.

12. Not enough Stone Age skeletons.

Evolutionary anthropologists now say that Homo sapiens existed for at least *185,000 years* before agriculture began,[28] during which time the world population of humans was roughly constant, between one and ten million. All that time they were burying their dead, often with artifacts. By that scenario, they would have buried at least eight billion bodies.[29] If the evolutionary time

scale is correct, buried bones should be able to last for much longer than 200,000 years, so many of the supposed eight billion stone age skeletons should still be around (and certainly the buried artifacts). Yet only a few thousand have been found. This implies that the Stone Age was much shorter than evolutionists think, perhaps only **a few hundred years** in many areas.

13. Agriculture is too recent.

The usual evolutionary picture has men existing as hunters and gatherers for 185,000 years during the Stone Age before discovering agriculture less than 10,000 years ago.[29] Yet the archaeological evidence shows that Stone Age men were as intelligent as we are. It is very improbable that none of the eight billion people mentioned in item 12 should discover that plants grow from seeds. It is more likely that men were without agriculture **for a very short time** after the Flood, if at all.[31]

14. History is too short.

According to evolutionists, Stone Age Homo sapiens existed for 190,000 years before beginning to make written records about **4,000 to 5,000 years** ago. Prehistoric man built megalithic monuments, made beautiful cave paintings, and kept records of lunar phases. Why would he wait two thousand centuries before using the same skills to record history? The biblical time scale is much more likely.

Dr. Humphries' information presents some challenges to an old earth, so who is right? Is the earth; 6,000 years old or 13.8 billion-years old? What if the answer to both questions is, YES?

Australian physicist Dr. Barry Setterfield may be able to support such a notion. Einstein suggested that "time is relative", but Setterfield takes this idea to the extreme by generating a bona fide, convincing set of arguments and theory that suggests the orbital age of the earth, the time it takes the earth to make a complete orbit of the Sun, vastly differs from its atomic age, which is measured by the rate of atomic delay. Without controversy, science agrees that every accurate measurement concerning the atomic age of the universe and the matter in it depends upon the speed of light being constant (C). Dr. Setterfield's theory is that the constant is not constant at all, but that the speed of light is slowing down and has done so since the beginning of the universe, in exponential proportions. Thus, the atomic age of the earth is not even close to its orbital age. In Setterfield's theory, billions of years of atomic time could fit nicely into thousands of orbital years. So, if the constant is not really constant, it is a new ballgame for time calculation and impacts the geological record and destroys the carbon dating scale.

The basis of his theory is based in two thoroughly explored astronomical principles, variations in both the 'red shift' and 'Plank's constant'. He claims that at some point in time, the earth stopped expanding rapidly, an idea rejected by the mainstream scientific theory, which proposes that the universe has expanded consistently for billions of years and continues to expand. The problem that Setterfield's theory creates is that most scientists use the rate of radioactive decay (C) of certain elements as a scale for determining elapsed time. All ordinary matter is made up of combinations of chemical elements, each with its own atomic number, indicating the number of protons in the atomic nucleus. Additionally, elements may exist in different isotopes, with each isotope of an element differing in the number of neutrons in the nucleus. A particular isotope of a particular element is called a nuclide. Some nuclides are inherently unstable. That is, at some point in time, an atom of such a nuclide will spontaneously transform into a different nuclide. This transformation may be accomplished in a number of different ways, including radioactive decay, either by emission of particles, usually electrons, positrons or by spontaneous fission, and electron capture.

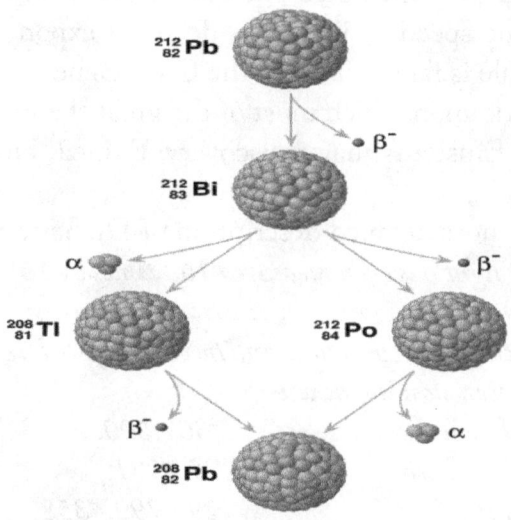

While the moment in time at which a particular nucleus decays is unpredictable, a collection of atoms of a radioactive nuclide decays exponentially at a rate described by a parameter known as the half-life, usually given in units of years when discussing dating techniques. After one half-life has elapsed, one half of the atoms of the nuclide in question will have decayed into a "daughter" nuclide or decay product. In many cases, the daughter nuclide itself is radioactive, resulting in a decay chain, eventually ending with the formation of a stable, nonradioactive daughter nuclide; each step in such a chain is characterized by a distinct half-life. In these cases, usually the half-life of interest in radiometric dating is the longest one in the chain, which is the rate-limiting factor in the ultimate transformation of the radioactive nuclide into its stable daughter. Isotopic systems that have been exploited for radiometric dating have half-lives ranging from only about 10 years for tritium to over 100 billion years for Samarium(147).

For most radioactive nuclides, the half-life depends solely on nuclear properties and is essentially a constant. It is not affected by external factors such as temperature, pressure, chemical environment, or presence of a magnetic or electric field. This predictability allows the relative abundances of related nuclides to be used as a clock to measure the time from the incorporation of the original nuclides into a material to the present. But, if the rate of decay, which is calculated directly from the speed of light, has declined exponentially through the ages, the scale is fatally flawed. The C in science is a sacred cow. It is the benchmark upon which most of the great theorems are based. It is the basis for Einstein's major discovery, E=mc2. How dare anyone assault it?

Below is a more detailed description of Dr. Setterfield's work:

The speed of light has been measured 163 times by 16 different methods over the past 300 years. However, Australian physicist Barry Setterfield and mathematician Trevor Norman, have suggested the speed of light appears to have been slowing down!

1657:	*Roemer*	**307,600.*
1875:	*Harvard*	*299,921.*
1983:	*NBS*	*299,792.4358*

*km/sec

Alan Montgomery's computer analysis supports the Setterfield/Norman results and indicates that the decay of velocity closely follows a cosecant-squared curve and has been asymptotic since 1958. If correct, the speed of light was 10–30% faster in the time of Christ; twice as fast in the days of Solomon; four times as fast in the days of Abraham, and perhaps more than 10 million times faster prior to 3000 BC.

A graph, from Setterfield's own website illustrates that the speed of light dropped very swiftly at first, as the ZPE increased just as rapidly, interfering with the travel of the photons, then more slowly as we approach the common age.

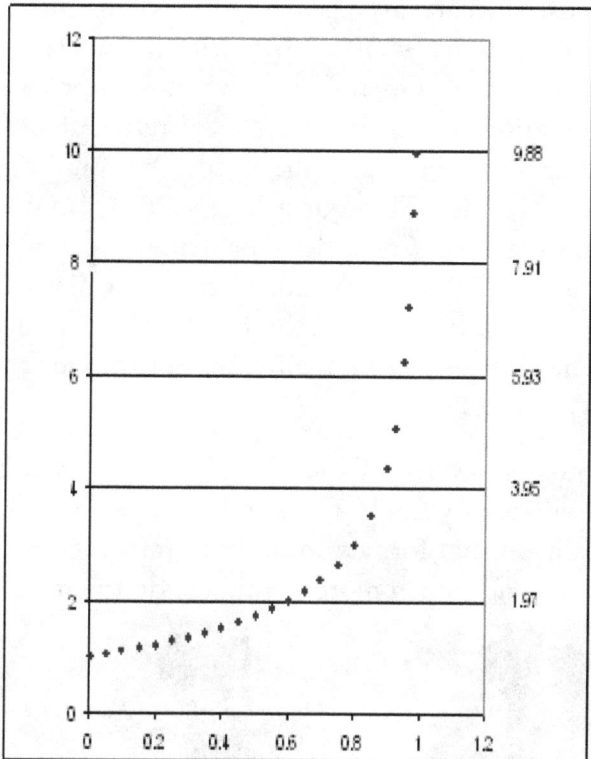

Creationists reason, this is why the sun, moon and stars are designated for keeping time (Genesis1:14). This is a deal changer for creationists, especially young earth creationists. Ken Ham is a staunch supporter of Setterfield's research and conclusions. Although this theory is extremely complex and requires some degree of expertise in physics, Setterfields lectures on YouTube give an exhaustive presentation of the theory and seem rational and reasonable. The fact that the relationship between the atomic year and the Solar year have never been explored speaks to a certain closemindedness, often displayed by the modern scientific community. Other of Setterfield's assertions, that the universe is no longer expanding rapidly, and that Planks constant is slowing also deserve more research and study.

Lastly, the Great Flood answers many of the evolutionist's questions about the age of the earth. Evolution and the great flood are incompatible; but the flood is essential for the creationist. Evolutionists are uniformitarian (millions of years of rarely interrupted bliss); creationists are catastrophists (regularly occurring catastrophes). Global catastrophes are vital to the science of creationism. This is especially true of the Great Flood, which compresses the earth's geological formation into a few short years rather than hundreds of millions of years and eliminates the need for evolutionists' millions of years story of geology. Flood or deluge stories, narratives in which a great flood, usually sent by a deity or deities, destroys civilization, often in an act of divine retribution, have been discovered in over 100 ancient cultures. In them all, the flood waters are described as a measure for the cleansing of humanity, in preparation for rebirth.

EXAMPLE 1 GEOLOGY

The photos below are of the Grand Canyon, which evolutions say it took millions of years to form and a canyon 1/140th the size of the Grand Canyon that formed in 38 hours after the Mt. St. Helens eruption. There is not much more to say, just do the math!

EXAMPLE 2 TOPOGRAPHY & COSMOLOGY

The Great Flood was cause by monsoon rains and that "the great fountains of the deep burst forth" (Genesis 7:11). These massive cisterns of superheated, but still liquid, water existed and still exist below earth's crust. In the violent separation of the continents (Pangaea); these fountains of the deep gushed out and massive volcanos erupted releasing steam and ash into the atmosphere. Pangaea is a well-supported scientific theory. Once again, the time of the separation is what is debated between creationists and evolutionists.

The fault lines and stretchmarks of the continental separation are clearly seen in the image below.

EXAMPLE 3 CLIMATE

Another indication of the Flood and when it happened are climate records that demonstrate a sudden temperature spike, followed by a rapid fall and gradual rebound around 10,000 BC. This exactly mirrors what the flood should have produced. The presence of a distinct cold period at the end of the LGM interval has been known as the Younger Dryas period or the *Bond Event #6, Yet, they have little evidence of what may have caused it. This is exactly what would have happened with a flood of global magnitude. Temperatures would have risen rapidly from volcanic and oceanographic activity, then suddenly plummeted with the formation of the polar ice caps.

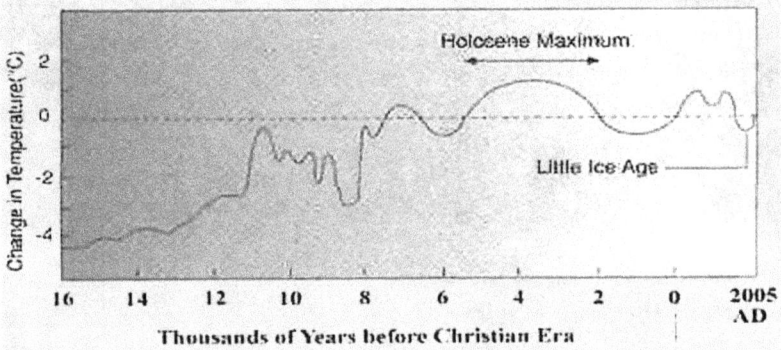

EXAMPLE 4 DEMOGRAPHICS

Another indication of a global catastrophe is that the accumulated mutations in the human genome are relatively few. Scientists attribute this to a microevolutionary bottleneck that occurred around 10,000 BC, when the human race came near to extinction and the entire human population was reduced to a very few individuals. To be expected, this reduced the genetic diversity of the population, then quickly rose after that.

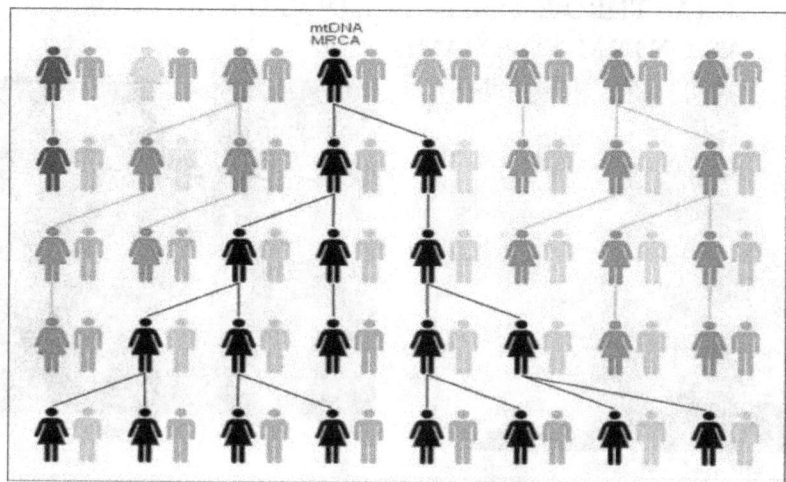

EXAMPLE 4 ANTHROPOLOGY – MICROEVOLUTIONARY BOTTLENECK

In addition, Grand civilization cannot be found before 10,000 BC. Evolutionary archaeologists agree that although intelligent humans (Homo sapiens) have existed for up to 300,000 thousand years, there is only evidence for sophistic culture as far back as 10,000 BC. No trace of agricultural development or settlements has survived longer than that. Building the Ark was an incredible feat. It required a sophistication in mathematics and engineering that is not seen until the Ziggurats and other pyramidal architecture of the early Bronze Age. That God revealed the mysteries of science to Noah is obvious. The Ark marked the beginning of technology that aided in the formation of the first great civilizations: Sumer, Egypt and the IVC. by the Mesopotamian Timeline.

EXAMPLE 4 ANTHROPOLOGY – THE DEVELOPMENT OF CIVILIZATION IN MESOPOTAMIA

There was an explosion of civilization in the Fertile Crescent, that corresponds with Genesis, chapters 6-10, established in the Mesopotamia Timeline, 10,000-2200 BC.

- 10000   End of the most recent glacial maximum
- 9000   Beginnings of agriculture in the Middle East
- 8000   Ovens in use in the Near East are applied to pottery production
- 7600   Counting Stones & Cylinder Seals in use throughout Mesopotamia
- 6600   Nineveh is first settled.
- 5200   Sumer inhabited by Ubaid people
- 5000   Irrigation begins in Mesopotamia
- 4900 – 1950 Sumerian civilization expands throughout the Tigris-Euphrates valley
- 4500   First walled city, Uruk in Mesopotamia.
- 4000   First settlement in Ur (home of Terra and Abraham)
- 3600   Invention of writing in Sumer
- 2750   The city of Tyre is founded
- 2700   Hatti (Hittites) establish trade with the city of Sumer
- 2400   Earliest Sumerian mention of migrating Amorites in Mesopotamia
- 2330   Sargon of Akkad sacks Ur, then the rest of Mesopotamia, creating the world's first empire

- 2200   Beginning of Bond Event #3; Abraham, Sarah, Terra and Lot leave Ur

EXAMPLE 5 GEOLOGY & TOPOGRAPHY – ZOOLOGY & BOTANY

The Global Megafauna Extinction of 10,000 BC is the final nail in the coffin. Paleontologists recognize the extinction of the mammalian megafauna as during the Holocene, 12,000 BP. Since the topography and climate of the Earth was radically altered after the Flood, many species, adapted to the predeluge world, did not survive. As our earth became colder, the vegetation changed, and animals that did not adapt rapidly died out. Climate change may also have driven human migrations from the Fertile Crescent, as game was reduced.

So far, we have looked at the age of the Earth through the lens of the physical sciences, Yet, there is biological and anthropological evidence that must be considered. Is there clear evidence of a time when intelligent, rational humans like us appeared on the Earth? Once again evolutionists say 100,000 years and six-day creationists 6,000 years. But, what do the facts say?

A recent scientific, genetic study of human genetics related to the brain concluded that a unique ASPM (Asporin) variant appeared 5000-15,000 years ago. This roughly correlates with the rapid development of written language, monumental architecture, refined agriculture and the consequential rise of civilization and cities. This lends support to the "Fortuitous Mutation" that Richard G. Klein refers to in his book, '*The Dawn of Human Culture*, in which he says this mutation occurred about 50,000 years ago.

Before six-day creationists throw this theory in the trash, forget the timing for now and remember that all we are saying here is that there is the possibility of the rapid improvement of mankind's ability to adapt to a more complex civilization. This could explain the explosion of technology in the days when Noah built the Ark or when Nimrod built the first great city, Babel. Could this phenomena correspond to the "Fortuitous Mutation" that Richard G. Klein refers to in his book, The Dawn of Human Culture?

A researcher at HHMI continues to examine this trail to real intelligence and claims that there is more modern evidence that suggests, as we would expect, that microevolution or adaptation continues today, especially as it relates to the human brain.

Lahn's analysis of genes indicates human brain continues to evolve

By Catherine Gianaro
Medical Center Public Affairs

Human evolution—in what has become our most important organ, the brain—is still under way, University researchers report in two related papers published in the Friday, Sept. 9 issue of *Science*. The studies show two genes linked to brain size are rapidly evolving in humans.

"Our studies indicate that the trend that is the defining characteristic of human evolution—the growth of brain size and complexity—is likely still going on," said lead researcher for both

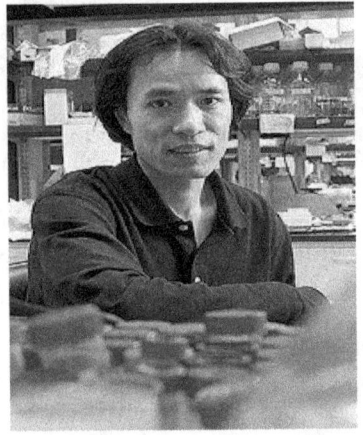

Bruce Lahn

papers Bruce Lahn, Assistant Professor in Human Genetics and an investigator in the Howard Hughes Medical Institute.

"Meanwhile, our environment and the skills we need to survive in it are changing faster than we ever imagined. I would expect the human brain, which has done well by us so far, would continue to adapt to those changes." Evolution, Lahn said, does not occur at the species level. Rather, some individuals first acquire a specific genetic mutation, and, because that variant confers on those who bear it a greater likelihood of survival, it then spreads in the population.

"We're seeing two examples of such a spread in progress," he said. "In each case, it's a spread of a new genetic variant in a gene that controls brain size. This variant is clearly favored by natural selection."

Lahn previously showed that there was accelerated evolution in humans among numerous genes, including microcephalin and ASPM (abnormal spindle-like microcephaly-associated). Both genes regulate brain size and therefore, said Lahn, "were good candidates to look for signatures of selection. We indeed found such signatures when we compared humans to other species," he said. "As a natural extension of that, we asked, could it be that selection on these genes is still ongoing in humans?"

In the two *Science* papers, the researchers looked at variations of microcephalin and ASPM within modern humans. They found evidence that the two genes have continued to evolve. For each gene, one class of variants has arisen recently and has been spreading rapidly because it is favored by selection.

For microcephalin, the new variant class emerged about 37,000 years ago and now shows up in about 70 percent of present-day humans. For ASPM, the new variant class arose about 5,800 years ago and now shows up in approximately 30 percent of today's humans. These time windows are extraordinarily short in evolutionary terms, indicating that the new variants were subject to very intense selection pressure that drove up their frequencies in a very brief period of time—both well after the emergence of modern humans about 200,000 years ago. Each variant emerged around the same time as the advent of "cultural" behaviors. The microcephalin variant appears along with the emergence of such traits as art and music, religious practices and sophisticated tool-making techniques, which date back to about 50,000 years ago. The ASPM variant coincides with the oldest-known civilization, Mesopotamia, which dates to 7000 BC.

"Microcephalin, has continued its trend of adaptive evolution beyond the emergence of anatomically modern humans. If selection indeed acted on a brain-related phenotype, there could be several possibilities, including brain size, cognition, personality, motor control or susceptibility to neurological/psychiatric diseases."

All this information fits neatly into universally accepted views of microevolution, or adaptation within individual species. So, it seems that the human skills necessary for adapting to civilization are continually improving. For creationists, these specific facts lead to the conclusion that a thoroughly modern man arrived in the form of Adam and Eve, who were created in God's image, with the ability to adapt to their environment. Evolutionists might say that this was the genetic breakthrough that differentiated Neanderthal or Denisovan man from modern man.

Before we proceed to anthropological, historical arguments, let us look at comments from respected historians and archeologists regarding the contributions of the Bible to the study of anthropology and ancient history.

Sir William Ramsay

Ramsay, an outstanding Near Eastern archeologists, says: "Luke is a historian of the first rank; not merely are his statements of fact trustworthy; he is possessed of the true historic sense; he fixes his mind on the idea and plan that rules in the evolution of history, and proportions the scale of his treatment to the importance of each incident. He seizes the important and critical events and shows their true nature at greater length...In short, this author should be placed among the very greatest of historians".

Jack Wellman

"Archaeological support for the Bible continues to swell, year after year. Today, well over 50,000 digs at 30,000 different locations have been excavated. These finds continue to provide substantial support of the Bible's written record. Some of the finds include Belshazzar, King Darius, and King Cyrus. What were once thought to be mythological empires mentioned in the Old Testament, like the Hittite's, have been found to have existed after all.

Since many unbiased scholars accept the Bible as an ancient historical source that has proven to be helpful to historians, anthropologists and archeologists, let's examine the Biblical record of the development of agriculture, previously mentioned in chapter one, which was the cornerstone to the rise of ancient civilizations, specifically Sumer, Egypt and the Indus Valley Culture.

The Bible repeatedly says that Adam and his immediate offspring were farmers.

Genesis 2:15 And the Lord God took the man and put him into the garden of Eden to dress it and too keep it."

Genesis 3:23 Therefore the Lord God sent him (Adam) forth from the garden of Eden to till the ground from whence he was taken."

Genesis 4:2 And Abel was a keeper of sheep, but Cain was a tiller of the ground (Adam's two sons).

So, can we make a case that the historical development of agriculture agrees with the Genesis narrative? The following article says that we can.

BIBLICAL RESEARCHBIBLICAL ARCHAEOLOGY

The Evidences for a Recent Dating for Adam, about 14,000 to 15,000 years Before Present

The great majority of the cultivated plants of the world trace their origin to Asia. Out of 640 important cultivated plants, about 500 originated in Southern Asia. Studies demonstrate that Asia is not only the home of the majority of modern cultivated plants, but also of our chief domesticated animals such as the cow, the yak, the buffalo, sheep, goat, horse, and pig.

More recent studies conducted by Melinda A. Zeder place the initial domestication of goats to the Zargos Mountains at about 10,000 years ago, pigs at 10,500 and cattle at 10,000. These dates mean that animals were domesticated at much the same time as crop plants, and bear on the issue of how this ensemble of new agricultural species — the farming package known as the Neolithic revolution — spread from the Near East to Europe. Studies indicate that large scale wheat cultivation began from 8,000 to 9,000 years ago in Mesopotamia.

Simcha Lev-Yadun, of Israel's Agricultural Research Organization, and colleagues suggest the first farms may have been between the Tigris and Euphrates rivers in what is today northeastern Turkey and northern Syria. Archaeological evidence shows that the earliest known farming settlements of the Fertile Crescent were in this core area. Also, the limited genetic variability of these crops implies that they were domesticated only once rather than by several different cultures at roughly the same time, about 9,300 years ago.

Genesis 11:2 And it came to pass, as they journeyed from the east, that they found a plain in the land of Shinar, and they dwelt there.

Archaeological evidence shows that around this time, farming techniques began to spread out of Anatolia, where the Ark is thought to have landed, across Europe and Asia. As already shown, the data on the farming indicates that the after the flood Genesis history took place in the Ararat area and that the area is also the origin of many of the known farm crops and domesticated animals.

As is always the case, climate variation had a dramatic effect on the spread and advancement of agriculture in Europe. Since 1916, with the onset and refinement of pollen analytical techniques, palynologists have concluded that the Younger Dryas represented a distinct period of vegetational change in large parts of Europe during which warmer climate vegetation was replaced by generally cold climate glacial plant succession. So, what would greatly increase the toil of a group of farmers more than a period of severe climate? So, it would seem that one could conjecture that the period of a relatively warm period of about 14,000 years ago was when Adam started farming. This warming trend was followed by a cool period of from 14,000 years ago to about 12,000, which could correspond to "the curse of the ground" spoken of in Genesis, chapter 3, a period in which farming was more difficult. Then about 12,000 years ago the warming begins again, and farming becomes easier and proliferates. The point is that six-day creationists do not rest their case in theology alone; their views have some basis in genuine academic thought as is shown in this comparison of the Biblical record with agricultural development.

Unfortunately, understanding 'the beginning', Creation, does not get much help from the original Old Testament language, Hebrew as the Hebrew word, *Yom*, usually translated as day, like our English word for day, can refer to a 24-hour day, sunrise to sunset or a long, unspecified period of time, as in "the day of the dinosaurs". In addition, the Hebrew word *ereb*, translated evening, also means "sunset," "night" or "ending of the day." The Hebrew word *boqer*, translated morning, also means "sunrise," "coming of light," "beginning of the day," or "dawning". Our English expression: "The dawning of an age" serves to illustrate this point. Likewise, this expression in Hebrew could use the word, *boqer*, for dawning, which, in Genesis 1, is often translated morning. The actual number of words in Hebrew is much fewer than that of the English translations. The words "and there was" are not included but were added to make the English flow better. The actual translation is "evening and morning 'n' day." There is no way to discern from the verbal context that the text is referring to 24-hour days.

The first thing one notices when looking at Genesis 1 is the unusual construction surrounding the words morning and evening together with day. This combination is very rare, occurring only ten times in the Old Testament, six of which, of course, are in the Genesis creation account. Not all instances of "morning" and evening" refer to a literal period of time? Here two specific examples:

1. *"In the <u>morning</u> it [grass] flourishes, and sprouts anew; toward <u>evening</u> it fades, and withers away." (Psalm 90:6)*

This verse refers to the life cycle of grass compared to the short life span of humans. Obviously, the grass does not grow up in one morning and die by the same evening. The period of time refers to its birth in the morning and its death in the evening, at least several weeks or months later.

> 2. *"And the vision of the <u>evenings</u> and <u>mornings</u> which has been told is true; but keep the vision secret, for it pertains to many <u>days</u> in the future."* (*Daniel 8:26*)

This verse actually refers to events that are yet to happen, which is currently 2500 years from the time when it was originally written. One could easily say that these mornings and evenings represent thousands of years. The apostle Peter tells us, "With God, a thousand years is as a day" (*2 Peter 3:8*). From a logical perspective, the third day of creation must have been longer than 24-hours. The text indicates a process that would take years. On this day, God allowed the land to produce vegetation and tress which bore fruit. The text specifically states that the land produced trees that bore fruit with seed in it. Any horticulturist knows that fruit-bearing trees require several years to grow to the maturity to be able to produce fruit. However, the text states that the land produced these trees (indicating a natural process) and that it all occurred on the third day. Obviously, such a "day" could not have been only 24 hours long.

From a purely rational standpoint, the events of the sixth day of creation, which is the only day for which we have exhaustive detail, seem to require more than a literal, 24-hour day. On this day, God created the mammals and mankind. The details of His careful fashioning of Adam from earthen clay indicates great care. Then he placed Adam in his new home, an expansive garden, gave him a tour, thoroughly instructed him on its care, and finally gave him specific warnings about the Tree od the Knowledge of Good and Evil. Next God brought all the species of animals, at least the family of genus, to Adam to be named. This job, in itself, could have taken many days, weeks, or months. After a suitable companion for Adam could not be found among the animals, God put Adam to sleep, took out a rib and created Eve. When Adam woke up, he used the Hebrew word pa'a⊠ma⊠h, which means "at long last." If Adam had only waited a few hours for his helpmate, it is unlikely he would have used this Hebrew word. The context suggests that Adam had to endure lonliness before Eve was created. So, it is very unlikely all of this could have taken place in 24 hours, especially since much of it was dependent upon Adam, who did not have the same abilities as God.

The belief that creation days are long periods of time is not just a recent interpretation of the scriptures, but was prevalent since the first century. Dr. Hugh Ross published a book in 1994 entitled <u>Creation and Time</u>, which documents in detail what first century Jewish scholars and the early Christian church fathers said regarding their interpretation of creation chronology. Jewish scholars include Philo and Josephus, while Christian Fathers include Justin Martyr, Irenaeus, Hippolytus, Clement, Origen, Lactantius, Victorinus, Methodius, Augustine, Eusebius, Basil, and Ambrose. Among this group, nearly all acknowledged the likelihood that the creation days were longer than 24 hours. The evidence presented in <u>Creation and Time</u> is both compelling and well documented. I highly suggest you read it.

Thorough investigation of this evidence, we conclude that 'the beginning' holds no answers to our inquiry. One way or another, it does seem clear from the evidence, that time, space energy, matter and the four fundamental forces, came into being suddenly and completely but the when remains a question. Since we all agree that matter changes but is neither created nor destroyed, the "Big Bang" seems to be a satisfactory name for what happened the beginning. The unanswered questions are whether the Big Bang was an intentional act of creative intelligence or is an unexplainable cosmic accident that mysteriously and unexplainable occurred 13.4-13.7 billion light years ago.

CHAPTER THREE
SEEKING ANSWERS FROM LIFE

If you think that the age of the universe is controversial, what about the age of and origin of life? Professor Stephen Hawking, one of the world's foremost cosmologists, put it this way in his documentary *Into the Universe; The Story of Everything* (2010): *"I am amazed at how much we understand about the origin of the universe, which occurred billions of years ago...but how little we know about the origin of life, which is far more recent."* He goes on to say that life remains a mystery, but our best guess is that it was simply a cosmic accident. Is this not an admission by the most brilliant mind of the 21st century that the question of the origin of life remains unanswered?

Another brilliant scientist and the vocal leader of the Darwinian evolutionary viewpoint, Richard Dawkins, said something similar in an interview recorded for the movie *'Expelled'*. In the absence of any cause or observable process for the creation of life, he admitted the possibility that life could have been 'seeded' on Earth by aliens, as is suggested by the Hollywood film "Prometheus". This is not what he really believes, but in the absence of clear evidence of spontaneous generation, he would rather admit that alien seeding is more possible than suggesting that it might have been a supernatural creative act.

So, why the confusion? All living creatures have some sort of DNA and are otherwise similar. Functions like respiration and reproduction are shared by all life forms. Yet, there are vivid contrasts, uniqueness and unexplained complexity in even the simplest life. This is no truer than with mankind. It is not just intelligence, language, opposable thumbs, conscience, creativity, self awareness, morality that make us unique, there is the intangible of conscious awareness. From the days of Plato and Aristotle insightful men have been analyzing the difference between us and the animals and there is universal acceptance that we are a distinct and special species. Creationists and evolutionists agree that species adaptation, also called micro evolution, natural selection, or survival of the fittest is good science. Micro evolution, the small, gradual changes within a species, which are beneficial for survival, cannot be denied. This phenomenon is scientifically observable, both at the physical and genetic level. Darwin is to be commended for his work in analyzing and defining the process of micro-evolution. Pretty good for a theologian.

Evidence is that humanity has adapted through the ages. There is no denying that we are vastly different from our ancestors and even from other humans living in different parts of the world today. Through the millennia, micro-evolution, or survival of the fittest, has helped us thrive as a species. If conditions demanded, we may have been more stooped, more muscular, more agile, more brutish or more dexterous, but we remain homo sapiens. The Zulus and the Pygmies, who live but hundreds of miles apart in Central Africa, represent the extremes of human variety. The Bushman are simple, nomadic, hunter-gatherers living mostly in the forest while the Massai, their neighbors, are mighty hunters, capable of incredible physical feats. Their physical size, strength, cunning, and athletic abilities clearly demonstrate adaptation.

Dr. Orville Boyd Jenkins writes,

"History indicates all humans in every place have a rich mix of genetic and cultural heritage. People and peoples are as we find them. We can only begin to probe that in various ways. There is a wide range of physical types among the peoples of Eastern Africa… the truth that the primary thing the various peoples of the Horn of Africa have in common is that they all live in Eastern Africa! This generally means that the population groups that now live in the African continent exhibit great diversity from one to the other. The focus is not on the variation within any one population group, but the variation among different population groups."

A 2012 article describes the complete genomic sequence of three African hunter gatherer populations, the Hadza, the Sandawe (both from Tanzania) and the Pygmies from Cameroon. The researchers were looking for telltale variants in the genetic code that could help explain differences between individuals and populations. Among them were sharp differences between the three groups in smell and taste - suggesting that "each population's senses had adapted to the new smells and foods they encountered."

The natural genetic variation within a population of organisms means that some individuals will survive more successfully than others within their environment. Factors which effect reproductive success are also important, an issue which Charles Darwin developed in his ideas on sexual selection. Natural selection acts on the phenotype, or the observable characteristics of an organism, but the genetic basis of any phenotype which gives a reproductive advantage will become more common in a specific population. Over time, this process can result in adaptations that specializes organisms for particular ecological niches; heat, cold, wet, dry, rich food source or poor food source. Evolutionists point out the size of the human brain pan as one important adaptation. An interesting fact about the brain pans of the supposed predecessors of humans, those that are credible, is that they are in many cases those discovered were as large, or in some cases larger, than those of modern man.

This largely unexplained anomaly is an issue you don't hear a lot about at least not in the mainstream but the answer to this puzzle might be vital for the equity of our journey, so let's open another Pandora's Box…

In 1993, Michael Cremo, a George Washington University graduate and former naval officer, co-wrote <u>Forbidden Archeology</u> with Richard L. Thompson.

Preview	*Forbidden Archeology: The Hidden History of the Human Race* **presents a representative sample of this anomalous evidence suggesting that humans have been on the earth for millions of years, just as the ancient Sanskrit writings of the Vedic literatures describe. The Vedic histories inform us that humans have existed since the beginning of the day of Brahma, about 2 billion years ago. Cremo and Thompson conclude that even the conventionally accepted evidence does not offer a cohesive picture of the *missing link*; instead, the multiplicity of proposed evolutionary linkages among the hominids in Africa creates a very confusing scheme of human evolution. They call for a drastic revision of the now-dominant assumptions about human origins.**

This book makes the point that modern man is not modern at all. In fact, it claims that humans like us have lived on the earth for millions of years, and that the scientific establishment has suppressed the fossil evidence of extreme human antiquity. He speaks about a knowledge filter as the cause of this suppression. Since Cremo is a Hindu, the book has attracted attention from some mainstream scholars as well as Hindu creationists. While some scholars of mainstream archeology and paleoanthropology have described the work as pseudoscience, some credible critics have acknowledged positive aspects of the book. Anthropologist Kenneth L. Feder wrote in his review:

"While decidedly antievolutionary in perspective, this work is not the ordinary variety of anti–evolutionism in form, content, or style. In distinction to the usual brand of such writing, the authors use original sources and the book is well written. Further,

the overall tone of the work is far superior to that exhibited in ordinary creationist literature."

Archaeologist Tim Murray also wrote a review of Forbidden Archeology:

"I have no doubt that there will be some who will read this book and profit from it. Certainly it provides the historian of archeology with a useful compendium of case studies in the history and sociology of scientific knowledge, which can be used to foster debate within archaeology about how to describe the epistemology of one's discipline." He also commented on the similarities in argument with those of Christian Creationists: "This is a piece of 'Creation Science' which, while not based on the need to promote a Christian alternative, manifests many of the same types of argument: first, an attempt to characterize the opposition as motivated by the need to preserve their view of the world rather than a desire to practice unfettered inquiry; secondly, to explain the currently marginal position of your alternative as being the result of prejudice, conspiracy and manipulation rather than of any fault of the theory itself; thirdly, to present the opposition, in this case mainstream palaeoanthropology and quarternary archaeology, as being united as a 'secret college' to manipulate the public mind and to exclude non-professionals from being able to control science for the benefit of all."

Cremo identifies himself as a "Vedic archeologist", since he believes his findings support the story of humanity described in the Vedas. The Indian magazine *Frontline* called Cremo and Thompson "the intellectual force driving Vedic creationism in America". Historians of science, Jo Wodak and David Oldroyd, published a 23-page review article, *Vedic Creationism: A Further Twist to the Evolution Debate in Social Studies of Science*. The review points out positive aspects of the book; that Forbidden Archeology brings to attention many interesting facts that have not received much consideration from historians; and the authors' detailed examination of the early literature is certainly stimulating and raises questions of considerable interest, both historically and from the perspective of practitioners of sociology of scientific knowledge. Cremo's books provided much of the content for the 1996 NBC special "*The Mysterious Origins of Man*".

Like Setterfield's model of time, Cremo's archeological evidence could be a game-changer in our quest for truth. It barges through strongly guarded doors and demands us to consider the implications of such a world. This gives new dimension to micro evolution and how mankind has become what he is today. It also might answer other unexplained questions from a mysterious time before antiquity that produced marvels like the Stones of Puma Punku, the Nazca Lines, the Piri Reis Map and the Antikythera Computer, but that is material for another book.

Yet, stretching Darwin's theory to the macro scale of species migration is unacceptable for creationists. Phillip Johnson goes to the heart of the matter:

> "Critics of evolutionary theory are well aware of the standard examples of microevolution, including dog breeding and the cyclical variations that have been seen in things like finch beaks and moth populations. The difference is that we interpret these observations as examples of the capacity of dogs and finches to vary within limits, not of a process capable of creating dogs and finches, much less the main groups of plants and animals, in the first place . . ."As any creationist (and many evolutionists) would see the matter, making the case for 'evolution' as a general theory of life's history requires a lot more than merely citing examples of small-scale variation. It requires showing how extremely complex biological structures can be built up from simple beginnings by natural processes, without the need for input or guidance from a supernatural Creator" (Reason in the Balance, 1995, p. 74).

Creationists do not believe Darwin's assertion that micro evolution is followed by macro evolution, the change of one separately distinct species to another separately distinct species. In other words, creationists believe that a wolf, *Canis lupus*, can become a bulldog, *Canis lupus familiaris*, but not a pig, *Sus scrofa domesticus*. A dog and a wolf are the same species because they can successfully breed and bear young that are able to reproduce. Whereas, dogs and pigs are two separately distinct species and cannot breed and produce offspring. Mules and a horse are separate but similar species that can breed and produce offspring, but that offspring is sterile, incapable of reproduction.

Evolutionists, both objectivist and theistic, believe that the entire process of macro evolution is observable science. Francis Collins, the man who unlocked the human genome, declares, "Evolution is no more a theory than that of a round earth". What confidence! The science of evolution does have scientific challanges, however. Creationists say that it may not be as 'cut and dried' as Dr. Collins supposes. DNA may be _God's Language_, but creationists might say that this language was written one species at a time. The detailed description of every living species is written with just four simple phosphate molecules, typically described as AGTC. That fact alone is used by the creationist as teleological proof for the existence of a Creator and is used by objectivists to support the theory of evolution.

Could an animal become thoroughly human over a vast evolutionary span? Could humans develop all of the characteristics necessary for survival through a developmental evolutionary process? Much has been written on the subject. In 1900, De Vries, Tschermack and Correns used Mendel's work to support Darwin's assertion that it could. Dennis Bonnette, the resident theologian for theistic evolution, writes: _"Evidence favoring the interconnectedness of species seems overwhelming... biochemical evidence, when mentioned with other information (the evolutionary path of Eohippus) has been so complete that we tend to dismiss any opposition to evolution."_

That being said, evolutionist bear the burden of proof. The theistic evolutionist faces the same challenge as does the objective evolutionist. Just believing in a Creator does not get one off the hook. Uncontroverted, empirical evidence for evolution must be demonstrated, since the similarity of genetic code could be explained by each species having the same designer. A Ford is recognizable as a Ford and a Rolex is recognizable a Rolex. There are distinguishing characteristics of each specific model, but there is also evidence of similar design and manufacturing.

Along these lines, much has been said and written about the genetic similarity of chimpanzees and humans. Up until recently, we were told by geneticists and anthropologists that the two species are 98.5% similar, but once again new, exculpatory evidence has arisen.

Human and Chimp DNA Only 70% Similar, According to This Study

February 22, 2013; by Jeffrey P. Tomkins

A chromosome-by-chromosome comparison of chimpanzee and human DNA. The chimp DNA was cut into slices of varying lengths (see legend on the right), and a similar sequence was searched for on the relevant human chromosome, which is shown on the horizontal axis.

*I have written about the similarity between human and chimpanzee DNA three times before. It's an important question for creationists, intelligent design advocates, and evolutionists alike, since the chimpanzee is supposed to be the closest living relative to human beings. As a result, a comparison of chimp DNA to human DNA gives us some idea of what the process of evolution would have to accomplish to turn a single apelike ancestor into two remarkably different species like chimpanzees and people. Early on, it was widely thought that human DNA and chimp DNA were 99% similar. As I discussed in my first post on this subject, that was based on a very limited analysis of only a minute fraction of human and chimp DNA. Now that the entire set of nuclear DNA (collectively called the "genome") of both humans and chimpanzees have been sequenced, we now know that the 99% number is just plain wrong. Interestingly enough, however, even though both genomes have been fully sequenced with a reasonable amount of accuracy, no one can agree on exactly how similar the two genomes are. Why is that? Because comparing genomes is **a lot harder** than you might think. While we know the **sequence** of the chimp and human genomes really well, we do not understand the DNA itself. Indeed, there are large sections* *of DNA that seem to be functional, but we simply have no idea what they do. As a result, comparing the genomes of two different species can be very, very tricky.*

> *"Genome-wide, only 70% of the chimpanzee DNA was similar to human under the most optimal sequence-slice conditions. While chimpanzees and humans share many localized protein-coding regions of high similarity, the overall extreme discontinuity between the two genomes defies evolutionary timescales and dogmatic presuppositions about a common ancestor. "*

The graph at the top of this article best shows his results. Notice that the similarity hovers around 70% for all chromosomes except the Y chromosome. The size of the "slice" affects the result a bit, but not much. In the end, this leads Dr. Tomkins to conclude: *Keywords*

- *author-jeffrey-tomkins*
- *dna*
- *human-evolution*
- *human-genome*

Featured In
- *Browse this issue*
- *Buy this issue*

Since the original 2005 report for the chimpanzee (chimp) genome assembly (5X rough draft), an additional one-fold redundant coverage has been added. Using the new 6X chimpanzee assembly, a sequential comparison to the human genome was performed on an individual chromosome basis. The chimpanzee chromosomes were sliced into new individual query files of varying string lengths and then queried against their human chromosome homolog using the BLASTN algorithm. Using this approach, queries could be optimized for each chromosome irrespective of gene/feature linear order. Non-DNA letters (gap filling 'N's) were stripped from the query data and excluded from the analyses. The definition of similarity for each chromosome was the amount (percent) of optimally aligned chimp DNA. This definition was considered to be conservative because it did not include the amount of human DNA absent in chimp nor did it include chimp DNA that was not aligned to the human genome assembly (unanchored sequence contigs).

For the chimp autosomes, the amount of optimally aligned DNA sequence provided similarities between 66 and 76%, depending on the chromosome. In general, the smaller and more gene-dense the chromosomes, the higher the DNA similarity—although there were several notable exceptions defying this trend. Only 69% of the chimpanzee X chromosome was similar to human and only 43% of the Y chromosome. Genome-wide, only 70% of the chimpanzee DNA was similar to human under the most optimal sequence-slice conditions. While, chimpanzees and humans share many localized protein-coding regions of high similarity, the overall extreme discontinuity between the two genomes defies evolutionary timescales and dogmatic presuppositions about a common ancestor.

Is this the last word on the subject? Most certainly not, but it is probably the best comparison attempt made so far. Also, the fact that the Y chromosome has a remarkably low level of similarity compared to the other chromosomes is consistent with other studies. In addition, as we learn more about genomes and how they work, we will probably find better ways to compare the genomes of different organisms. For right now, however, it seems clear that humans and chimpanzees are not nearly as genetically similar as most evolutionists would have us believe.Technical This study should remind us that creationists and IDT scientists have been asking for more physical evidence for macro evolution; 'the irrefutable proof', if you will. Although evolutionists have long said that the fossil record is evidence, Creationists say the evidence shows an explosion of living things, the Cambrian Explosion, ordered as separate, distinguishable species over a short period of time. They would say that this is the smoking gun that kills Darwin's theory of macro evolution and neutralizes evolutionists' claims.

It might surprise you to learn that many Christians are theistic evolutionists and accept the Genesis narrative as a sort of allegorical, understandable way to interpret the evolutionary process, but choosing to accept that position demands supporting evidence. Furthermore, it requires an explanation of how a dumb animal became human. Two Popes, Pius XII and John Paul II, and one ecumenical council made room for Roman Catholics to accept theistic evolution. Unfortunately, for those who go down this pathway, the Popes and the council did not remove responsibility for 'keeping the sacredness of the gospel intact', and therein lies the issue. For the Christian Gospel to be valid, numerous conditions relating to Adam, Eve and Jesus Christ must be clearly established. Catholics, who would be theistic evolutionists and genuinely believe that God's plan for the redemption of mankind is vital and earnestly desire to uphold the important theological conditions revealed in scripture, must grapple with these conditions.

They begin with Adam being in the essential theological state of being made "in the image of God". Then, the theology of redemption or soteriology continues with Eve coming directly from Adam ("bone of my bone and flesh of my flesh"). Then, both Adam and Eve must have lived in a state of innocence and then willfully sinned against God to receive the justly pronounced curse of death, both spiritual and physical. They must also have been cursed with a sin nature (the knowledge of good and evil) that was passed to their progeny. Lastly, every human who has ever lived must have descended directly from Adam and Eve since Christ, the Second Adam, must have descended from Adam and Eve, so he could represent the entire human race when he paid for our sins on the cross. Augustine explains that Christ could not redeem the angels because they were created individually, not as a race. He had only one life to give, so he gave it to purchase the pardon for every human being.

With each of these elements in place, Jesus Christ, the seed of the woman described in Genesis 3:16, as direct descendent of Adam (Son of Man), could represent the entire human race on the cross of redemption just as Adam represented our entire race in the fall. As one of the first Popes, the Apostle Paul, put it "for as in Adam all die so in Christ shall all be made alive", Romans 5:18. I know that for many of you this means nothing, but for Christians who decide to be theistic evolutionists, it places a burden of immense consequence. How do you reconcile these specific and critical requirements with evolution? We will examine this more fully in the next chapter.

Finally, it must be stated that the mere anthropological proof of the existence of humanoids like mankind does not alone prove evolution. There are published theories that propose that it possible that there was a time when the earth was inhabited by three or more separate and distinct humanoid species that never shared an evolutionary link. David Pilbeam, in his article "Human Origins and Evolution" (Ed. by A.C. Fabian, Cambridge, 1988, p.111) says of human development, commencing from Homo Erectus, the first hominid species to achieve a wide distribution outside Africa:

> *"H. Erectus and its industries lasted from over 1.5 my ago to less than 0.5 my ago without much change, implying a degree of behavioral stability which was surprising when it first became apparent to paleoanthropologists. Around 0.3 my ago, hominids with somewhat larger brains (1100-1300 cm3) and slightly differently shaped skulls began to appear: so-called archaic Homo Sapiens; Overall, they were still very similar to Homo Erectus. To distinguish them from us, the Modern Homo Sapiens, the archaic modifier is added. The best known archaic sapiens came from Europe and West Asia: the Neanderthals. However, despite their sapient label, despite attempts to 'launder' them, I do not think they should be called Homo Sapiens because morphologically, and by inference, behaviorally they differ markedly from modern humans."*

Pilbeam goes on to say (pp.112-3):

> *"A substantial evolutionary change comes between archaic and modern Homo Sapiens. It occurred over the period between 130 ty and 30 ty ago, ending with the appearance of humans like us, with less massive skeletons, less muscular bodies, rounder heads and flatter faces. Judging from their archeological traces, their behavioral potential was also like ours by at least 35 ty ago. The cause of this last major change in human development is obscure, and there is considerable disagreement about its tempo and mode, when and where it happened and what micro evolutionary pattern was followed. With these modern humans we get the first evidence for recognizably modern behavior patterns of many kinds; from cave painting and sculpture, to sophisticated tools and elaborately planned co-operative hunting, to increased population size and*

density and broadened geographical and ecological range. It is here, finally, that all the truly 'human' characters come together for the first time."

The Neanderthals are recognized as a subspecies of Homo Sapiens who first appeared around 100,000 or less years ago. There seems little doubt that Cro-Magnon man or what are termed early modern humans and Neanderthals co-existed on the planet. According to Sarah Bonney, writing in *New Scientist*, January 20, 1990, "Skeletal remains provide good evidence of early modern humans in Israel (Qatzeh and Skhul Caves) between 100,000 and 90,000 years ago. By at least 43,000 years ago, populations of modern people had reached southern and central Europe." The *New Scientist*, issues of November 26, 1987 and February 25, 1988, held that in the Middle East, the two forms of Neanderthal and Cro-Magnon may have co-existed for tens of thousands of years.

According to James Bischoff, from the use of new dating techniques, it appears that early modern humans reached the Spanish L'Arbreda Cave in Catalonia and El Castillo Cave in Cantabria around 40,000 years ago (*Journal of Archeological Science*, Vol 16, pp.563 and 577). Both caves had previously been used by Neanderthals. Neanderthal tool technology is known as Mousterian, Cro-Magnon is known as Aurignacian; and where skeletal evidence is absent the tool types are taken as evidence of type. Regardless of the academic debate and the absolute accuracy of the dating, this all suggests that multiple humanoid species may have existed on the planet prior to the genesis narrative of creation. According to Bunney, "In southwestern France and along the northernmost coast of Spain, as in the Middle East, there are signs that modern humans and Neanderthals lived side by side for a time."

The existence of three separate species, without any evidence of intermediary mutants or changes, does not argue that one evolved from the other; indeed, it argues the reverse. The absence of mutant change in Homo Erectus from 1.5-.5 million years ago and the emergence of a distinct species 300,000 years ago, in the case of the Neanderthals, and 35-40,000 or more years ago, in the case of Cro-Magnon, with no mutants either differing or deleterious being found indicates a stable genetic structure. with a secondary structure imposed on the first.

It would appear that the major factor showing the marked divergence of hominid development was between that of H. Erectus and H. Neanderthalis. Erectus had a narrow pelvis and reduced brain case. The reduced pelvis indicates a much reduced gestation period and hence a smaller, less developed infant and brain capacity. Despite all the anthropological argument, there does not appear to be any reason why Erectus should be viewed as human in any sense other than using sticks and stones for tools, a characteristic also common to apes and sea otters.

The change from H. Erectus to H. Neanderthalis, however, is an enormous genetic change with the restructuring of gestation, of sexual behavior and hence familial groupings. Neanderthals had brain capacity equal to ourselves and a better muscled, stronger body. Because of a rounder face for colder climates according to some anthropologists, a case has been developed for reduced frontal lobes. Earlier arguments were for a greater brain size than us, although these may be isolated to some specimens.

Neanderthals were late-appearing humanoids separate to and unlike anything preceding them, and, contrary to popular evolutionary theory, there is no evidence and no logical reason for accepting that they were the antecedents of the modern humans. Modern DNA testing has revealed that they are utterly unrelated to humans having a 27 strand DNA system, giving us more in common with rats than them. There is no logically compelling reason why one should accept that Homo Sapiens or Homo Neanderthalis evolved from any other species, nor is there any evidence that indicated that such a construction occurred.

It seems that Dr. Hawking is correct...the existence of life generates more questions than concrete answers! Life seems to be mysteriously silent about its origin. We would all agree that it is of similar design and that a good case can be made for macro evolutionary development. Yet, there seems to be an enormous body of evidence to support the claims of objectivists, compatibilists, theists and literalists and just as many contradictions. Although this may seem like 'strike three', in our journey for truth, let us call this one a foul ball, so we are not 'out'! Let's continue to seek answers from the experts. If anyone knows, they should.

CHAPTER FOUR
SEEKING ANSWERS FROM THE EXPERTS

I t was no trouble finding expert opinions form every camp. They write books, make videos, record documentaries and sell product. The creation-evolution controversy fuels them and their related businesses and gives them a platform from which to rail and rail they do!

The designated spokesman and entrepreneurial genius for the literal, 24-hour day creationist movement is **Ken Ham**. He is the founder and President of *Answers in Genesis*, a think tank of scientists and theologians who aggressively defend the literal six-day position. The organization includes the Creation Museum and a Noah's Ark Adventure near Cincinnati, OH, which attract thousands of visitors annually and a book publishing operation. Ken conducts a series of regional conferences each year, usually held in large churches. He also hosts *Answers. . . with Ken Ham*, a 60-second program broadcast daily on radio stations and the Internet featuring Ham's commentary on the issues of creation.

Having attended Ham's conferences, I can tell you that his arguments are well received by the conservative, Christian Community. He is a hero to those who need to be literalists because, like Ken, they believe that any compromise related to the literal interpretation of Genesis is a threat to the literal interpretation of the rest of the Bible. Ham's staff and associates are convincing authors and speakers, with eminent qualifications and experience. Much of Ham's work is built on the foundation of the renown theologian, Henry M. Morris, author of <u>The Genesis Record.</u>

Ham proposes that the Universe was created about 6,000 years ago, and that Noah's flood occurred around 2500 BC. Much of the dating on which Ham depends is taken from the work of Bishop James Usher (1581-1656), primate of Ireland. Usher developed a complete chronology of human history, based on genealogical and historical records from the Bible. Ussher's chronology represented a considerable feat of scholarship: it demanded great depth of learning in what was then known of ancient history, including the rise of the Persians, Greeks and Romans, as well as expertise in the Bible, biblical languages, astronomy, ancient calendars and chronology. Ussher's account of historical events for which he had multiple sources other than the Bible is usually in close agreement with other, more modern accounts. For example, he placed the death of Alexander in 323 BC and that of Julius Caesar in 44 BC.

Faced with inconsistent texts of the Torah, each with a different number of years between Flood and Creation, Ussher chose the Masoretic text rather than the Septuagint, which claims an unbroken history of careful transcription stretching back centuries. His choice was confirmed in his own mind, because it placed Creation exactly four thousand years before 4 BC, the generally accepted date for the birth of the 'Jesus' of the Christian faith. Moreover, he calculated that Solomon's temple was completed 3,000 years after creation, so that there were exactly 1000 years from the temple to Jesus, who was supposedly the human embodiment of the material Temple.

Ham contends that the animals carried on Noah's ark produced the biological diversity observed on Earth today and that dinosaurs co-existed with modern humans in the pre-flood world. He supports his view with scripture and credible evidence on display at the Creation Museum. Ham accepts that natural selection can give rise to a number of variations from an original population, by Mendelian recombination of already existing genes, but that new genes cannot arise from mutations, because this would require adding genetic information. He claims that only intelligence can cause a beneficial mutation, that mutations and natural selection can only remove preexisting information and that species are of the same *kind*, and no new *kind* can arise from this process.

Ham questions the reliability of radiometric dating; a technique used to date objects such as moon rocks, fossils and human artifacts. Since 1989, Ham has frequently asked the question, "Were you there?" regarding the origins of life and evolution, implying that knowledge of unwitnessed events can only be inferential and not observational. He continues, "Creationists and evolutionists, Christians and non-Christians, all have the same evidence—the same facts. Think about it: we all have the same earth, the same fossil layers, the same animals and plants, the same stars—the facts are all the same. The difference is in the way we all interpret the facts."

So why do we interpret facts differently? Because we start with different presuppositions; these are things that are assumed to be true without being able to prove them. These then become the basis for other conclusions. All reasoning is based on presuppositions (also called axioms). This becomes especially relevant when dealing with past events." Many evolutionists and even some creationists take issue with Ken Ham. Yet, he is convinced that science cannot prove that God did not create the universe in six days. His immovable position only reinforces the disagreement that exist between Naturalists and Creationists and even OECs and young earth YECs.

Another defender of the Creationist viewpoint is **Paul Nelson**. After receiving his B.A. in philosophy with a minor in evolutionary biology from the University of Pittsburgh, Nelson gained a Ph.D. in philosophy of biology and evolutionary theory from the University of Chicago. Paul A. Nelson has been involved in the international intelligent design debate for three decades. He is currently a Fellow of the Discovery Institute and Adjunct Professor in the Master of Arts Program in Science & Religion at Biola University.

Nelson was an associate of Professor Phillip Johnson of UC-Berkeley, author of the bestselling critique, Darwin on Trial. Nelson was an organizer of the Mere Creation conference (1996), where the modern intelligent design research community first formed. Nelson's research interests include the relationship between developmental biology and our knowledge of the history of life, the theory of intelligent design, and the interaction of science and theology. Nelson's scholarly articles have appeared in journals such as Biology & Philosophy, Zygon, Rhetoric and Public Affairs, and Touchstone, and chapters in the anthologies Mere Creation (Intervarsity Press), Signs of Intelligence (Brazos), Intelligent Design Creationism and Its Critics (MIT Press), and Darwin, Design, and Public Education (Michigan State University Press).

Nelson is a fellow of the Discovery Institute's Center for Science and Culture and of the International Society for Complexity, Information and Design. The Discovery Institute is a U.S. non-profit public policy think tank based in Seattle, Washington, best known for its advocacy of intelligent design. Its "Teach the Controversy" campaign aims to teach creationist, anti-evolution beliefs in United States high schools, positing a scientific controversy exists over these subjects. Thus far, Nelson and his associates have been prohibited from teaching ID creationism in the public schools by the evolutionary scientific community and the courts. In an interview for *Touchstone Magazine* Nelson said that the main challenge facing the ID community is to "develop a full-fledged theory of biological design".

Quite literally, in the other corner is the bold and unusually aggressive leader of the modern objectivists, **Richard Dawkins**, Ph.D. An evolutionary biologist, Dawkins is beyond certain that all living things appeared through evolution, which began around 4 billion years ago. Dawkins came to prominence with his 1976 book The Selfish Gene, which popularized the gene-centered view of evolution. In his 2006 book The God Delusion, Dawkins contends that a supernatural creator almost certainly does not exist and that religious faith is a delusion—"a fixed false belief" or consonant dissonance as previously discussed.

In a set of controversies over the mechanisms and interpretation of evolution, in what has been called 'The Darwin Wars', one faction is often named after Dawkins, while the other faction is named after the late American paleontologist Stephen Jay Gould, reflecting the pre-eminence of each as a popularizer of the pertinent ideas. In particular, Dawkins and Gould have been prominent commentators in the controversy over sociobiology and evolutionary psychology.

Dawkins is a prominent critic of creationism. He has described the Young Earth creationist view, that the Earth is only a few thousand years old, as "a preposterous, mind-shrinking falsehood", and his 1986 book, The Blind Watchmaker, contains a sustained critique of the argument from design, an important creationist argument. In the book, Dawkins argues against the watchmaker analogy made famous by the 18th-century English theologian William Paley via his book Natural Theology, in which Paley argues that just as a watch is too complicated and too functional to have sprung into existence merely by accident, so too must all living things—with their far greater complexity—be purposefully designed. Michael Behe has also contributed significantly to this particular debate.

Dawkins shares the view generally held by objectivists, that natural selection is sufficient to explain the apparent functionality and non-random complexity of the biological world. In 1986, Dawkins and biologist John Maynard Smith participated in an Oxford Union debate against A. E. Wilder-Smith, a Young Earth creationist and Edgar Andrews, president of the Biblical Creation Society. In general, however, Dawkins has followed the advice of his late colleague Stephen Jay Gould and refused to participate in formal debates with creationists because "what they seek is the oxygen of respectability", and doing so would "give them this oxygen by the mere act of *engaging* with them at all". He suggests that creationists "don't mind being beaten in an argument. What matters is that we give them recognition by bothering to argue with them in public." There is much debate over who prevailed in these debates.

In a December 2004 interview with American journalist Bill Moyers, Dawkins said that "among the things that science does know, evolution is about as certain as anything we know." When Moyers questioned him on the use of the word *theory*, Dawkins stated that "evolution has been observed. It's just that it hasn't been observed while it's happening." He added that "it is rather like a detective coming on a murder after the scene... the detective hasn't actually seen the murder take place, of course. But what you do see is a massive clue... huge quantities of circumstantial evidence." Dawkins has ardently opposed the inclusion of intelligent design in science education, describing it as "not a scientific argument at all, but a religious one". He has been referred to in the media as 'Darwin's Rottweiler', a reference to English biologist T. H. Huxley, who was known as 'Darwin's Bulldog'.

Much of Dawkins' specific attacks have focused on **Michael Behe**, one of the leaders of the IDC movement. An American biochemist and author, Behe currently serves as professor of biochemistry at Lehigh University in Bethlehem, Pennsylvania, and as a senior fellow of the Discovery Institute's Center for Science and Culture. Behe is best known for his argument for irreducible complexity, which asserts that some biochemical structures are too complex to be adequately explained by known evolutionary mechanisms and are therefore more probably the result of intelligent design. Often called the mouse trap argument, Behe uses the interconnected workings of the independent parts of a mouse trap to demonstrate that the evolution of complex structures requires the independent development of parts that are useless and not beneficial until the entire structure is complete. Dawkins strongly contests this argument, because many now consider it a major impediment to evolutionary theory.

Behe has testified in several court cases related to intelligent design, including the court case *Kitzmiller v. Dover Area School District* that resulted in a ruling that intelligent design was religious in nature. Behe's claims about the irreducible complexity of essential cellular structures have predictably been rejected by the vast majority of the evolutionist scientific community but remains a major plank in the creationist platform. Behe says he once fully accepted the scientific theory of evolution, but that after reading *Evolution: A Theory In*

Crisis, by Michael Denton, he came to question evolution. Later, Behe came to believe that there was evidence, at a biochemical level, that there were systems that were irreducibly complex. These were systems that he thought could not, even in principle, have evolved by natural selection, and thus must have been created by an intelligent designer, which he believed to be the only possible alternative explanation for such complex structures.

In March 1992, a conference at Southern Methodist University brought Behe together with other leading figures to develop a strategy to allow ID to be taught in public schools along with evolution. This strategy was later called the "wedge strategy". Following a summer 1995 conference called 'The Death of Materialism and the Renewal of Culture,' the group obtained funding through the Discovery Institute. In 1996 Behe became a senior fellow of the Discovery Institute's Center for the Renewal of Science and Culture, later renamed the Center for Science and Culture, dedicated to promoting intelligent design. In 1993, Behe wrote a chapter on blood clotting in Of Pandas and People, presenting arguments which he later presented in very similar terms in a chapter in his 1996 book Darwin's Black Box. Behe later agreed that they were essentially the same when he defended intelligent design at the Dover Trial. Behe's refusal to identify the nature of any proposed intelligent designer frustrates evolutionists, who would love to move him from his scientific box to a religious one, thus making him an easier target.

Francis Collins, project manager for the Human Genome Project, makes a sincere and passionate attempt to insert God into Dawkins' arguments. An evangelical Christian and theistic evolutionist, Collins maintains that life had to have a designer and controlling hand in order to produce the diversity we observe, especially in the case of mankind. He takes particular exception to the idea that life could arise from non-living matter. I quote, "no current hypothesis comes close to explaining how that in the space of a mere 150 million years, the prebiotic environment that existed on planet Earth gave rise to life." In honesty and humility Collins refuses to admit what he knows is not true.

To be fair to Collins, it is notable that in the Ben Stein film 'Expelled', under relentless questioning in an interview, Richard Dawkins seemed lost for an explanation for the emergence of life from inanimate matter. He reluctantly and regrettably admitted that it might that life might have been sown on earth by extraterrestrials. Hawking, Collins and Dawkins, three of the greatest intellects of our time, are honest enough to admit that science's inability to create life in the laboratory gives rise to an intellectual void for evolutionary theory and deals a death blow to proving that life was a cosmic accident, emerging from a primordial pool of ooze.

Yet, Collins remains a strong critic of creationism. In his book, The Language of God, he presents a scathing attack upon creationist and IDC leaders by accusing them of manipulating science to fit their preconceived model. He also challenges creationist ideas by saying, "one would expect that that the rank and file of working biologists would also show interest in pursuing these (IDC) issues, especially since a significant number of biologists are believers." This might be true in a neutral environment, but Collins' assertion was put to rest in the movie 'Expelled", which showed that one after another science professionals were shunned, refused tenure and even fired for voicing support for just exploring IDC ideas. Isn't it amazing that in a society in which half of the people surveyed express belief in a Creator, the whole of our government institutions, public universities, publishers, and the media are bent on finding, outing and destroying people who simply want to investigate scientific ideas, other than evolution?

What is amazing is that these 'experts' work with the same set of facts but always draw dissimilar scientific conclusions about creation. Behe looks at the irreducible complexity of bacterial flagellum, sort of outboard motors which help bacteria navigate, and discerns that none of the 29 mutations that might have caused the evolution of that appendage were at all beneficial until every one of them was in place. Dawkins sees a potential benefit in each mutation, regardless of the inability of the flagellum to function until all were in place. Collins sees 'fundamentalist Christians' as ignoring and manipulating science to their benefit, and he and the rest of the 'enlightened evangelicals', who might have been called Gnostics at one time, are the only true guardians of orthodox science. One side sees the possible discovery of life on Mars as the ultimate proof of spontaneous generation; the other says that it would prove nothing.

Speaking of Mars, it is the hottest topic in science today. The Red Planet brings even more disagreement between these experts. Before this book is published, it is anticipated that life may be discovered on Mars. The degree of complexity will be the surprise but even finding bacteria will be earth-shaking news. Of course, each group will once again find different interpretation in the same facts. Evolutionists are giddy over the possibilities:

> *"If life on Mars is ever proven to exist (or have existed at some point in time), it would mean that the creation of life is not something that happens because of freak chance or divine influence, but is in fact a probable occurrence given the right conditions. Even further, if all that life requires is an aqueous solution like liquid water to grow and thrive (which is the current theory), then the universe is literally teeming with life. The suspected liquid water oceans on some of Jupiter's moons (Europa and Callisto) could be filled with life, and life could still be present underneath the Martian surface, where liquid water and thermal energy are still present." (from* www.marsnews.com)

The Institute for Creation Research, however, finds the comments of Kenneth Nealson, a NASA microbiologist, comforting and contradictory to the objectivist assumptions.

> *"Furthermore, if scientists were to find clear, unmistakable fossilized bacteria in a meteorite (or soil) from Mars in the future, it doesn't necessarily mean such bacteria evolved from non-life over millions of years. Applying uniformitarian assumptions, secular scientists conclude: "Just as the Mars meteorite . . . is thought to have been tainted by Earthly bacteria, samples from Mars, too, may not be what they seem." "We think there's about 7 million tons of earth soil sitting on Mars," says Kenneth Nealson, a microbiologist with NASA. "You have to consider the possibility*

that if we find life on Mars, it could have come from Earth."
(Newsweek, September 21, 1998, p. 12.)

NASA

Once again with Mars, the same facts but dissimilar conclusions. Missing the step of observation keeps all our experts in the realm of theory, ever discovering and always interpreting the evidence with a bias towards their own philosophy. It is divisive disagreement; each camp holding their 'own truth' about science and creation. We theologians call this eisegesis, studying the scriptures with a predisposition of what it means. The proper method, exegesis, allows scripture, taken in textual, historical and cultural context, to say what it says, free of preconceptions. Make no mistake, scientists, like theologians, are not free from predisposition. Their entire life forms a point of reference from which they cannot escape.

There is no doubt that our so-called experts are experts, they are experts in defending the particular philosophy that best interprets their view of science. I do not know about you, but I feel a little frustration setting in. Might we find the answers we seek in a more thorough study of anthropology?

CHAPTER FIVE

SEEKING ANSWERS FROM ANTHROPOLOGY

For objectivists, theists and theistic evolutionists, anthropology is the key to determining when modern mankind arrived on the scene. All three viewpoints agree that humans are not dumb animals, operating only on instinct. Mankind is obviously separate and distinct from the rest of animal kingdom. Did we evolve into such higher beings or were we created as such from the start? The answer is a quandary, and initiates five other foundational inquiries for the theistic evolutionist:

1. At what point did primates become human beings?
2. How did morality, intellect, judgment, compassion and spirituality take hold of a particular pre-human species?
3. Was there an actual couple, an 'Adam and Eve', who made such a transition together, simultaneously?
4. How might such 'woke' human beings have survived among other non-human species?
5. Was innocence a reality? If so, when and how was it shattered?

These abstract questions do not concern the objectivist or the creationist. The creationist's version of the direct creation of Adam and Eve by Almighty God avoids such nuance as do the atheistic or agnostic evolutionist's, who do not care. It is theistic evolutionists who have a 'dog in this hunt'. It is Rand's curse of 'eating your cake and having it too'. If one believes in a Creator and a creation story, he/she must find a practical way to explain the part of the Genesis narrative that says that the Creator fashioned man from the dirt with his own hands and breathed life into Adam's lifeless lips.

Do anthropologists and philosophers, most specifically those who support theistic evolution, have responsible answers to these complex questions? You bet they do! Dennis Bonnette, a Doctor of Philosophy from Notre Dame, and an orthodox Roman Catholic professor at Niagara University is an avid theistic evolutionist. Following the Church's allowance to consider theistic evolution, as long as the Gospel message remains intact, Bonnette provided convincing, scholarly answers to these questions in his book, <u>Origin of the Human Species</u> (Sapentia Press, 2007). It is a laborious and masterfully laid out presentation, which I highly recommend for those who want to delve more deeply into an orthodox alignment and underlying philosophy for theistic evolution. Dr. Francis Collins, in his book, <u>The Language of God</u>, makes a similar, but less comprehensive, philosophical argument.

According to Dr. Bonnette, the evolutionary tree of the 21st century is a little less cluttered at the top, as mainstream evolutionists and like-minded anthropologists propose that the earliest documented members of the genus Homo are Homo habilis, the earliest species for which there is positive evidence of use of stone tools, evolving around 2.3 million years ago. A million or so years later, Homo erectus, the first hominid to use controlled fire, arrived on the scene. Homo sapiens, anatomically correct modern humans, evolved possibly from Homo heidelbergensis, some 50,000 to 100,000 years ago, replacing local populations of Homo erectus. Since they say a picture is worth a thousand words, here is the creative art for what evolutionists propose.

Now, pick the first human. Which one stopped being a brute and became like you and me? Was it the use of tools that pushed mankind beyond the animal barrier? Was it the use of fire? Was it speech? Would the specific moment of intellectual enlightenment have been recognizable and distinguishable? For Roman Catholic theistic evolutionists, it must be.

Tools are used by many members of the animal kingdom. Otters select rocks from the ocean floor and river bottoms with which they relentlessly and skillfully crush clams. Chimpanzees use long reeds to extract termites from various environments for food. Tracing the development of tools used by hominoids requires more specific investigation. Evolutionists think that the earliest humanoid tools were made 3.3 million years ago. These prehistoric tools were made from stone, wood and animal bone, antlers and sinew.

Controlling fire certainly takes more intelligence, more focus, and more skill. All evidence of control of fire during the Lower Paleolithic is uncertain and has at best limited scholarly support. In fact, definitive evidence of controlled use of fire is one of the factors characteristic of the transition from the Lower to the Middle Paleolithic in the period of 400,000 to 200,000 BC. East African sites, such as Chesowanja near Lake Baringo, Koobi Fora, and Olorgesailie in Kenya, show evidence that fire was utilized by early humans from one million BC. Koobi Fora sites show evidence of control of fire by Homo erectus, with the reddening of sediment that can only come from heating at 200–400 °C. A hearth-like depression exists at a site in Olorgesailie, Kenya. In Gadeb, Ethiopia, fragments of welded tuff that appeared to have been burned were found among created Acheulean artifacts. A site at Bnot Ya'akov Bridge, Israel, has been claimed to show that *Homo. Erectus* made fires between 790,000 and 690,000 BC.

More conservative anthropologists state that archeological evidence suggests that cooking fires began in earnest only 250,000 years ago, when ancient hearths, earth ovens, burnt animal bones, and flint appear across Europe and the Middle East. Two million years ago, the only sign of fire is burnt earth with human remains, which most other anthropologists consider to be mere coincidence rather than evidence of intentional fire. Paleoanthropologist C. Loring Brace agrees with Allman's theory that cooking did not affect development, stating that he has only found evidence of earth-oven cookware from the past 200,000 years, which does not correlate with the earliest known use of fire from 800,000 years ago.

Why is the control of fire important? Wrangham suggested that by cooking meat, it acted as a form of "pre-digestion", allowing less food energy intake to be spent on digesting the tougher proteins such as collagen and complex carbohydrates. Suzana Herculano-Houzel calculated that if we ate only raw, unprocessed food, humans would need to eat for 9.3 hours per day in order to fuel their brains, which use about twice as much resting energy by percentage as other primates. Neurobiologist John Allman states that cooking food contributed to the development of the brain in early modern humans. Carel van Schaik believes that cooking did contribute to human evolution and Leslie

Aiello suggests, that numerous things contributed to the growth of the brain; increased consumption of meat, a smaller digestive system, cooking, and walking upright. Most anthropologists accept that the first humans made and used tools and controlled fire, but there is one more essential that we must consider; speech.

Since the early 1990s professional linguists, archaeologists, psychologists,anthropologists, and others have attempted to address, with newer methods, what they are beginning to consider "the hardest problem in science". Let's see what they have to say. Approaches to the origin of language can be divided into two groups, according to their underlying assumptions.

1. Continuity theories are based on the idea that language is so complex that one cannot imagine it simply appearing from nothing in its final form. Thus, it must have evolved from earlier pre-linguistic systems among our primate ancestors.
2. Discontinuity theories are based on the opposite idea, that language is a unique trait so it cannot be compared to anything found among non-humans and must therefore have appeared fairly-suddenly during the course of human evolution.

Another contrast is between theories that view language mostly as an innate faculty that is largely genetically encoded, and those that view it as a system that is mainly cultural, learned through social interaction. Evolutionists assume the development of primitive language-like systems as early as Homo habilis, while others place the development of symbolic communication only with Homo erectus or even Homo heidelbergensis, 600,000 years ago and the development of language proper with *Homo sapiens* less than 200,000 years ago.

Noam Chomsky , a distinguished proponent of 'discontinuity theory', is a renowned expert concerning the nature of innate universal grammar. He argues that a single chance mutation occurred in a single human about 100,000 years ago, triggering the instantaneous emergence of language in perfect or near-perfect form, a sort of linguistic punctuated equilibrium. Evolutionists assumed that the only change needed was the cognitive ability to construct and process recursive data structures in the mind. From these assertions should we assume that this instantaneous genetic alteration of the human language faculty is logical? Only if we do, can we accept his theory that language did appear rather suddenly within the course of human evolution.

In 2014, MIT made the following announcement:

Researchers from MIT and several European universities have shown that the human version of a gene called Foxp2 makes it easier to transform new experiences into routine procedures. When they engineered mice to express humanized Foxp2, the mice learned to run a maze much more quickly than normal mice. The findings suggest that Foxp2 may help humans with a key component of learning language — "transforming experiences, such as hearing the word 'glass' when we are shown a glass of water, into a nearly automatic association of that word with objects that look and function like glasses," says Ann Graybiel, an MIT Institute Professor, member of MIT's McGovern Institute for Brain Research, and a senior author of the study.

This research seems to confirm Professor Chomsky's assertions, but Professor Suzanne Kemmer proposes that speech may not be as simple as a single mutation. She suggests that speech in humans bridges cognitive, social, and physical dimensions:

Three Dimensions of Development in the History of the Human Species: Neuro-Cognitive, Social, and Physical

....

From Inquiry: Critical Thinking Across the Disciplines, Winter, 1996. Vol. XVI, No. 2.
By Linda Elder

Some Preliminary Distinctions

I shall argue that critical thinking cannot successfully direct our beliefs and actions unless it continually assesses not simply our cognitive abilities, but also our feeling or emotion states, as well as our implicit and explicit drives and agendas. I shall argue, in other words, that critical thinking provides the crucial link between intelligence and emotions in the "emotionally intelligent" person. Critical thinking, I believe, is the only plausible vehicle by means of which we could bring intelligence to bear upon our emotional life. It is critical thinking I shall argue, and critical thinking alone, which enables us to take active command of not only our thoughts, but our feelings, emotions, and desires as well. It is critical thinking which provides us with the mental tools needed to explicitly understand how reasoning works, and how those tools can be used to take command of what we think, feel, desire, and do. Through critical thinking, as I understand it, we acquire a means of assessing and upgrading our ability to judge well.

Two Contrary Tendencies of the Human Mind

Every human being enters the world with an initial motivation to have its way and to get what it wants, and thus "naturally" sees the world as designed to cater to its desires. This fact is apparent when we observe the behavior of young children. Their unfailing motto: "It's mine!" As we grow older, we learn methods for getting our way, which are much less blatant and thus less obvious to the untrained eye. Throughout our lives, our own desires and narrow interests are typically in the foreground of our thinking.

As we mature, we learn multiple ways to manipulate others, to influence or control others to get what we want. We even learn how to deceive ourselves as to the egocentrism of our behavior. We have no difficulty coming to conceptualize ourselves as fair-minded, empathetic, kind, generous, thoughtful, and considerate, as concerned, in short, with other persons.

The result is a kind of dualism in us: our selfish, egocentric side, on the one hand, and our capacity to recognize higher values on the other.

Tendencies Toward Rationality

Although we often approach the world through irrational, egocentric tendencies, we are also capable, as I have suggested, of developing a "higher" sense of identity. We are capable of becoming non-egocentric people, both intellectually and "morally." Science itself exists only because of the capacity of humans thinking in a non-egocentric fashion--intellectually speaking. Moral concepts, in turn, exist, only because of the human capacity to conceive of responsibilities that by their very nature presuppose a transcendence of a narrow moral egocentrism.

At a minimum, then, I envision the human mind as utilizing its three basic functions (thought, feeling, and desire) as tools of either egocentric or non-egocentric tendencies, both intellectually and morally.

From making fairly sophisticated tools, to cooking with fire, to verbal communication, through cognitive understanding and finally with the big leap to emotional intelligence, we must all agree that the jump from animal to human is huge. Maybe Dr. Bonette and others were too quick to call homo erectus a thoroughly modern human. Maybe it came much later. Once again, the evidence is conflicting. Yet, at some point evolutionists, both theistic and atheistic, agree that modern humans evolved. That just begins the real process of understanding from where morality, compassion, spirituality, accountability and all other uniquely, distinctive human characteristics came. Some would call this the soul.

The human soul, is a subject deeper than the Laurentian Abyss, so where do we begin? Maybe by looking at mankind's own beliefs about the soul. The soul, in many mythological, religious, philosophical, and psychological traditions, is the incorporeal and, in many conceptions, immortal essence of a person, living thing, or object. According to some religions including the Abrahamic religions in most of their forms, the immortal human soul, is capable of union with the Creator and is unique to mankind. The Catholic theologian Thomas Aquinas attributed soul (anima) to all organisms, but taught that only human souls are immortal. He further understood the soul to be the first actuality of the living body.

Other religions teach that all biological organisms have souls, and that even non-biological entities such as rivers and mountains possess souls. This latter belief is called animism. The term Soul can function as a synonym for spirit, mind, psyche or self. The words soul and psyche can also be treated synonymously, although psyche has more physical connotations, whereas soul is connected more closely to spirituality and religion. Although the terms soul and spirit are sometimes used interchangeably, the soul may denote a worldlier and less transcendent aspect of a person. According to psychologist James Hillman, the soul has an affinity for negative thoughts and images, whereas spirit seeks to rise above the entanglements of life and death. Since creationists believe that God breathed the spirit into mankind, he may be right.

Let us examine the soul from a historical perspective:

Plato

Drawing on the words of his teacher Socrates, (469 BC – 399 BC), Plato considered the soul the essence of a person, being that which decides how we behave. He considered this essence to be an incorporeal and eternal. In death, the soul is continually reborn in new bodies. The Platonic soul comprises three parts:

1. the logos, or logistikon (mind, nous, or reason)
2. the thymos, or thumetikon (emotion, or spiritedness)
3. the eros, or epithumetikon (appetitive, or desire)

Each of these has a function in a balanced, level, and peaceful soul.

Aristotle

Plato's overachieving student Aristotle, the tutor of Alexander the Great, defined the soul or psyche as the first actuality of a naturally organized body. Yet he argued against it having a separate existence from the physical body. In Aristotle's view, the primary activity of a living thing constitutes its soul. The various faculties of the soul or psyche, such as nutrition, sensation, movement, and so forth, when exercised, constitute the second actuality, or fulfillment, of the capacity to be alive. A good example is someone who falls asleep, as opposed to someone who falls dead; the former actuality can wake up and go about their life, while the second actuality can no longer do so. Aristotle identified three hierarchical levels of living things: plants, animals, and people. Reason is unique to humans alone.

Eastern Philosophers

Following Aristotle, the eastern philosophers Avicenna Ibn Sina and Ibn al-Nafis, further elaborated on the Aristotelian understanding of the soul and developed their own theories. They both made a distinction between the soul and the spirit, and in particular, the Avicennian Doctrine on the nature of the soul was influential among the Scholastics. Some of Avicenna's views on the soul included the idea that the immortality of the soul is a consequence of its nature. In prison, Avicenna wrote his famous 'Floating Man' thought experiment to demonstrate human self-awareness and the substantiality of the soul. He concluded that the idea of self is not logically dependent on any physical thing, and should be considered as a substance. This argument was later refined and simplified by René Descartes when he stated: "I can abstract from the supposition of all external things, but not from the supposition of my own consciousness."

Kant

In his discussions of rational psychology, Immanuel Kant (1724–1804) identified the soul as the "Id" and that the existence of an inner experience can neither be proved nor disproved. "We cannot prove the immateriality of the soul, but rather only so much: that all properties and actions of the soul cannot be cognized from materiality." It is from the soul that Kant proposes transcendental rationalization but cautions that such rationalization can only determine the limits of knowledge if it is to remain practical.

Hillman

Contemporary psychology is defined as the study of mental processes and behavior. However, the word psychology" literally means study of the soul, and psychologist James Hillman, the founder of archetypal psychology, has been credited with restoring the soul to its psychological sense. Although the words soul and spirit are often viewed as synonyms, Hillman argues that they can refer to antagonistic components of a person. Hillman's psychology is an attempt to give attention to the oft-neglected soul, which Hillman views as the "self-sustaining and imagining substrate upon which consciousness rests". Hillman described the soul as that "which makes meaning possible, [deepens] events into experiences, is communicated in love, and has a religious concern, as well as a special relation with death". Hillman takes the Neoplatonic stance that there is a third, middle position in which soul resides.

Contemporary Philosophers

The soul has less explanatory power in a western worldview, which prefers is more evident in the brain. Yet, philosophers such as Thomas Nagel and David Chalmers have argued that the correlation between physical-brain states and mental states is not strong enough to support identity theory. Nagel argues that no amount of physical data is sufficient to provide the 'what it is like' of first-person experience, and Chalmers argues for an explanatory gap between functions of the brain and phenomenal experience. On the whole, brain/mind identity theory does poorly in accounting for mental phenomena of qualia and intentionality. While neuroscience has done much to illuminate the functioning of the brain, much of subjective experience remains mysterious.

Creationists

For the sake of understanding each others' position, let's look at the soul from a Creationist, both Christian and Islamic, perspective. Most Christians understand the soul as an ontological reality distinct from, yet integrally connected with the body. Dualists believe that the soul and spirit are one, but there are tripartists who argue that they are distinct and inseparable (Hebrews 4:12). Its characteristics may be described in moral, spiritual, and philosophical terms. According to orthodox Christian eschatology, when people die their souls survive.

Among Christians, there is considerable debate regarding the point at which the fetus acquires a soul and consciousness. This is the reasoning behind the Christian belief that abortion should not be legal. Since most Christians regard the soul as the immortal essence of a human; the seat of human will, understanding, and personality, they reason that a fetus is a living human being. Augustine, one of western Christianity's most influential early Christian thinkers, described the soul as "a special substance, endowed with reason, adapted to rule the body". As mentioned previously, most evangelical Christians espouse a trichotomic view of humans (I Thessalonians 5:23), which characterizes humans as consisting of a body (soma), soul (psyche), and spirit (pneuma).

Philosopher Anthony Quinton states that the soul is a series of mental states connected by continuity of character and memory, and is the essential constituent of personality. The soul, therefore, is not only logically distinct from any particular human body with which it is associated; it is also what a person is. Souls have sensations and thoughts, desires, and beliefs, and perform intentional actions, so are essential. Richard Swinburne, a Christian philosopher of religion at Oxford University, wrote that it is a frequent criticism of substance dualism that dualists cannot say what souls are.

The origin of the soul has been hotly debated in historical theology. The major theories include soul creationism, traducianism and pre-existence. According to creationism, each individual soul is created directly by God, at the moment of conception. Traducianism presupposes that the soul comes from the parents by natural generation, the same as the body. According to the preexistence theory, souls pre-exist and are placed in the zygote at the moment of conception.

The Catechism of the Catholic Church defines the soul as "the innermost aspect of humans that which is of greatest value in them, that by which they are most especially in God's image: the soul signifies the spiritual principle in man." The Catholic Church teaches that the existence of each individual soul is dependent wholly upon God: "The doctrine of the faith affirms that the spiritual and immortal soul is created immediately by God." Eastern Orthodox and Oriental

Orthodox views are somewhat like Roman Catholic views, although different in some specifics.

Protestants generally believe in the soul's existence, but fall into two major camps about what this means in terms of an afterlife. Some, following Calvin, believe in the immortality of the soul and conscious existence after death, while others, following Luther, believe in the mortality of the soul and unconscious 'sleep' until the resurrection of the dead. Other Christians reject the idea of the immortality of the soul. They consider the soul to be the life force, which ends in death and will be restored in the resurrection. Theologian Frederick Buechner sums up this position in his 1973 book Whistling in the Dark: "We go to our graves as dead as a doornail and are given our lives back again by God, just as we were given them by God in the first place".

Similarly, John Thomas, founder of Biblical Unitarianism, was convinced that we are all created out of the dust of the earth and became living souls once we received the breath of life based on the Genesis account of humanity's creation. Adam was said to have become a living soul. His body did not contain a soul, rather his body plus the breath of life together were called a soul, in other words a living being. He proposes that we are mortal and when we die our breath leaves our body, and our bodies return to the soil. We are mortal until the resurrection from the dead when Christ returns to this earth and grants immortality to the faithful. In the meantime, the dead lie in the earth in the sleep of death until Jesus returns. This belief is sometimes referred to as "soul sleep".

Islam takes a much more mysterious approach to the soul. According to the Quran, Ruh (Spirit) is a command from Allah (God). And they ask you, O Muhammad, about the soul (Rûh). Say, "The soul (Rûh) is of the affair of my Lord. And mankind have not been given of knowledge except a little." [Quran 17:85]

Yet, like most of Christianity, Islam teaches the soul is immortal and eternal. What a person does is definitely recorded and will be judged at the utterly court of the God. From the Holy Quran Chapter 39 Surah Zumar verse 42:"It is Allah that takes the souls at death: and those that die not (He takes their souls) during their sleep: those on whom He has passed the Decree of death He keeps back (their souls from returning to their bodies); but the rest He sends (their souls back to their bodies) for a term appointed. Verily in this are Signs for those who contemplate."

It seems that the soul these philosophers describe, both by objectivist and theist, is unique to mankind; therefore, it must have an inception. When did the theistic evolutionist's dumb animal receive a human soul? According to Dennis Bonnette, the soul arrived in the package that included all the pieces that made us human rather than animal. Bonnette's first human or 'Adam' is the theistic evolutionists link to God's plan for the redemption of lost mankind. I must say that his explanation of Eve is less convincing. Yet, Dr. Bonnet gives theistic evolution a leg on which to stand and attain the Pope's challenge. It is only through the preservation of this vital part of soteriology that Bible-believing Christians can consider theistic evolution.

CHAPTER SIX

SEEKING ANSWERS FROM COPERNICUS, COLUMBUS, AND GALILEO

During the 15th – 17th centuries, great scientists, explorers, and thinkers walked where we are walking today, facing similar challenges as are discussed in this book. Without sophisticated equipment or government funding and facing the horrors of the Inquisition, they postulated ideas like the orbit of the planets around the Sun and the possibilities of travel across an undiscovered planet. Even thinking such things was taboo, much less attempting to establish them as truth. They were similar to Michael Cremo and Barry Setterfield, proposing new ideas in a closed off scientific community, that has already made up its mind.

COPERNICUS

Although previously studied by Hypatia of Alexandria in the 4th century, in 1543 Nicolaus Copernicus, a Polish astronomer and committed Christian, published his treatise *"De Revolutionibus Orbium Coelestium"* (On the Revolutions of the Heavenly Spheres). He dared to present a heliocentric model of the universe. Yet, even though his observations and mathematical calculations were correct, it took about 200 years for the heliocentric model to replace the Ptolemaic model. It should be noted that Ptolemy was a pagan Greek.

Thomas Kuhn's comment on the slow response to Copernicus' obvious genius is relevant to our discussion:

"To describe the innovation initiated by Copernicus as the simple interchange of the position of the earth and sun is to make a molehill out of a promontory in the development of human thought. If Copernicus' proposal had had no consequences outside astronomy, it would have been neither so long delayed nor so strenuously resisted".

As is the case today, Kuhn indicates that the philosophies of that day: Platonism, Roman Catholicism, Judaism, Islam, Deism and a host of other beliefs, overshadowed the pure scientific discovery, making progress tedious.

BRAHE & KEPLER

The Danish astronomer Tycho Brahe, who proposed a compromise between the geocentric and the heliocentric theories with his own Tychonic System, contributed to the 'revolution' by showing that the heavenly spheres were at best mathematical devices rather than physical objects. He reasoned that since the great comet of 1577 passed through the spheres of several planets, and the spheres of Mars and the Sun passed through each other, their physical nature is subordinate. Brahe and his assistants made numerous and painstaking observations which allowed Johannes Kepler to derive his laws of planetary motion. Kepler proposed an alternative model in 1605, essentially the one finally proven beyond a doubt, in which the planetary orbits were ellipses, rather than circles modified by epicycles as Copernicus used. Kepler's revised heliocentric system gave a far more accurate description of planetary motions than the Ptolemaic one.

GALILEO

Starting with his first use of the telescope for astronomical observations in 1610, Galileo Galilee provided support for the Copernican system by observing the phases of Venus and the moons of Jupiter, which showed that the apparently anomalous orbit of the Moon in Copernicus' theory was not unique. Galileo also wrote a defense of the heliocentric system, *"Dialogue Concerning the Two Chief World Systems"* in 1632, which led to his trial and house arrest by the Inquisition.

Throughout the same period, a number of writers inspired by Copernicus, such as Thomas Digges and Giordano Bruno, who argued for an infinite or at least indefinitely extended universe, with other stars as distant suns. Although initially opposed by Copernicus and Kepler, by the middle of the 17th century this became widely accepted, partly due to the support of René Descartes.

NEWTON

The Copernican revolution was completed by Isaac Newton, another outspoken Christian, who's *"Philosophiae Naturalis Principia Mathematica"* (1687) provided a consistent physical explanation which showed that the planets are kept in their orbits by the familiar force of gravity. Newton was able to derive Kepler's laws as good approximations and to get yet more accurate predictions by taking account of the gravitational interaction between the planets. Yet, the Roman Catholic Church seemed determined to retain the Geocentric model since they saw mankind as the crowning accomplishment of creation and the center of the universe.

COLUMBUS

Likewise, Christopher Columbus, known as the boldest navigator of his day, faced a related, equally challenging issue. Was the Earth a flat plane or a sphere? The Roman Catholic Church said it was the former, his observations said the later. So, could the Earth be successfully navigated? The models of the solar system developed by the Copernicans, asserted that the Earth, Sun and planets were orbiting spheres. Columbus had observed tall ships sailing to port near his home in Genoa. The tops of the sails always appeared first followed by the masts and then the deck. Based on this observation, he sailed west and the rest is history. Unlike the astronomers, he actually risked his life to make a scientific observation.

These brave men and women took risks; they faced opposition, as does anyone with a bold new theory. Hypatia was murdered for daring to be different. Some were summoned to appear before the Inquisition. It is said that Columbus only escaped imprisonment by summoning an Old Testament scripture to his defense. The text is Isaiah 40:22, which states that God "sits above the circle of the Earth". This one thing might have saved Christopher Columbus from being convicted as a heretic and never discovering the New World.

Why are new ideas suppressed? Because men are stubborn, we cling to old ideas like our comfortable bed shoes. We live in the past and enjoy our own comfort zones, but eventually true science triumphs or at least we hope that it does. Unfortunately, this is still the case today and those with their heels dug deep in the sand live on both sides of the evolution/creation debate. Scientists like Dr. Barry Setterfield and Michael Cremo, who have fact-based theories that challenge everything we know about the universe, have not received a fair hearing. The cool conservative men on both sides had already begun to formulate their critical responses before the ink dried on their revolutionary theories. Rather than being discussed, their ideas were discarded and dismissed! Yet, in a hundred years or so, their proposals may prove to be as authentic as those of past heroes.

Let's face it; the biggest obstacle in the case of Copernicus, Galileo and Columbus was the scientific ignorance and disinformation of their day. They had to face the archaic belief that the Earth's shape is a plane or disk, long before the church existed. Many ancient cultures have had conceptions of a flat Earth, including Greece until the classical period, the Bronze Age and Iron Age civilizations of the Near East until the Hellenistic period, India until the Gupta period and China until the 17th century. It was also typically held in the aboriginal cultures of the Americas, and a flat Earth domed by the firmament in the shape of an inverted bowl is common in pre-scientific societies.

Jewish acceptance of a flat earth is found in biblical and post-biblical times. Unfortunately, the Roman Catholic Church was in a bully position, as well. Using obscure, poetic scriptures, not designed to be taken literally, they had made the greatest of hermeneutical errors; taking scripture out of context. Revelation 7:1 is sometimes cited, which speaks of "four angels standing at the four corners of the earth." However, this passage makes reference to the cardinal directions as seen on a compass; north, south, east and west. Another passage often referred to is:

Psalm 75:3: *"When the earth and all its people quake, it is I who hold its pillars firm."*

Today, freshman in a good Bible college understand that there is a difference between Biblical poetry and history, parable and narrative. An understanding of 'context' is fundamental to interpretation, so why did men of much greater learning make such a huge mistake? The answer is obvious to church historians. Their preconceived notions were mostly Aristotelian, the founder of western thought. This was the principle issue and major cause of the Protestant Reformation; the church was more prone to tradition than scripture alone (sola scriptura).

The fact is there are numerous Scriptures that indicate that the Earth is a sphere. These scriptures were penned centuries before Copernicus, some before Aristarchus, an ancient Greek astronomer and mathematician who presented the first known model that placed the Sun at the center of the known universe, with the Earth revolving around it. He was influenced by Philolaus of Croton and was able put the other planets in their correct order of distance around the Sun. His astronomical ideas were often rejected in favor of the geocentric theories of Aristotle and Ptolemy.

Job 26:7 explains that the earth is suspended in space, the obvious comparison being with the spherical sun and moon. It says, "He stretcheth out the north over the empty place, *and* hangeth the earth upon nothing." Jesus said "For as Jonah was three days and three nights in the belly of the great fish, so will the Son of Man be three days and three nights in the heart of the earth"; indicating a spherical Earth with a center. Luke 17:34-36 also implies a spherical earth. Jesus said that at His return some would be asleep at night while others would be working at day time activities in the field. This is a clear indication of a spherical revolving earth, with day and night occurring simultaneously.

Yet, from the time of the Romanization of the Catholic or 'universal' Church, under Constantine, a new idea began to develop. Whereas the church had always followed scripture alone (sola scriptura) as a guide for doctrine and practice, now apostolic tradition was taking hold as an alternate guideline. Thus, mainstream Christianity moved away from the perceived simplicity of scripture to something more malleable, changeable, and formational, the supposed habits and beliefs of godly men and Aristotle.

Had this not been the case during the lives of our science heroes, things might have been different. Thus, it was the dogma of Romanism, not the scriptures themselves that cause the problem which science and exploration faced. Nonetheless, the damage was done. Christianity was dealt a body blow, from which in the eyes of many, it can never recover. Thus, the derogatory term, 'flat-earther', for those who believe creationism in any form. The truth is many of the great scientific minds of the 16th-19th centuries were devout Christians: Pascal, Mendel, Newton, Boyle, Pasture and Kelvin. Each was able to reconcile scripture and science, and none accepted a flat Earth position and their view of Darwinian evolution was skeptical, at best.

Looking at how this controversy played out, is it any wonder that the clergy in the 21st century struggle with similar issues that our Christian-scientist heroes of astronomy and exploration from the past faced? This is clearly demonstrated in a recent survey conducted by the Barna Group, which was commissioned by BioLogos: https://biologos. org/blogs/archive/a-survey-of-clergy-and-their-views-on-origins

CLERGY VIEWS: CREATION AND EVOLUTION

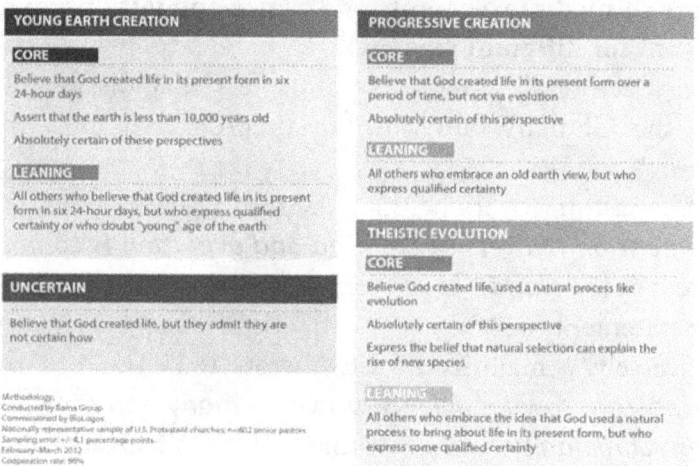

YOUNG EARTH CREATION

CORE

Believe that God created life in its present form in six 24-hour days

Assert that the earth is less than 10,000 years old

Absolutely certain of these perspectives

LEANING

All others who believe that God created life in its present form in six 24-hour days, but who express qualified certainty or who doubt "young" age of the earth

UNCERTAIN

Believe that God created life, but they admit they are not certain how

Methodology:
Conducted by Barna Group
Commissioned by BioLogos
Nationally representative sample of U.S. Protestant churches; n=602 senior pastors
Sampling error +/- 4.1 percentage points
February–March 2012
Cooperation rate: 95%

PROGRESSIVE CREATION

CORE

Believe that God created life in its present form over a period of time, but not via evolution

Absolutely certain of this perspective

LEANING

All others who embrace an old earth view, but who express qualified certainty

THEISTIC EVOLUTION

CORE

Believe God created life, used a natural process like evolution

Absolutely certain of this perspective

Express the belief that natural selection can explain the rise of new species

LEANING

All others who embrace the idea that God used a natural process to bring about life in its present form, but who express some qualified certainty

Overall, while a slight majority of the pastors surveyed fall under the label of Young Earth Creationism (54%), sizeable portions of clergy accept Progressive Creation (15%) and Theistic Evolution (18%).The numbers varied widely based on a number of factors, however. Pastors of mainline churches were most likely to accept Theistic Evolution, while non-Mainline, Charismatic, and Southern Baptist pastors were overwhelmingly Young Earth Creationists. Pastors of larger churches were also more likely to accept Theistic Evolution. Regionally, the highest percentage of YEC pastors was found in South, while the highest percentage of pastors accepting TE was in the Midwest. Pastors from the western states were the least likely to accept TE.

#2: Most pastors think science and faith questions are important.

Regardless of their views, the majority of pastors surveyed feel that the Church needs to look at how it handles issues of science. 72% of pastors with YEC views and 73% of pastors with TE views agree with the statement that *"the Christian community needs to take a serious look at its understanding of science and human origins in order to maintain its witness in the world."* (The numbers are slightly lower for pastors who hold to Progressive Creation and who are uncertain).

Similarly, 66% of YEC pastors and 61% of both TE and Progressive Creation pastors agree that *"younger adults today are more concerned than ever about whether faith and science are compatible."*

#3: Clergy think disagreements on science and faith harm our witness (but for different reasons).

Clergy across all three viewpoints feel that disagreements are harming the Church's outreach, but they differ in how they view that harm.

YEC pastors overwhelming agreed (85%) that *"Christian disagreement on matters of creation and evolution is compromising our witness to the world."* However, a majority of TE pastors disagreed with the statement (63%).

Conversely, a majority of TE pastors (63%) agreed that *"The church's posture toward science prevents many non-Christians from accepting Christianity,"* while a majority of YEC and Progressive Creation leaning pastors disagreed (59%).

#4: Pastors aren't avoiding science.

The majority of pastors think that addressing issues of science for their congregations is an important part of their work. Of those surveyed, 72% felt that addressing science issues in the local community was somewhat (51%) or very (21%) urgent. When asked about science on a national and global level, even more pastors felt that addressing science issues is important (43% somewhat and 46% very). Furthermore, 79% of pastors included scientific themes in at least one sermon in the past year, and 40% had included them in at least ten sermons.

The majority of clergy across all four viewpoints also agreed with the statement *"Just as scripture should influence human interpretation of science, science should also inform our understanding of scripture."* The numbers were highest for TE pastors and those who are uncertain (81% and 72%, respectively), though over half of YEC and PC pastors also agreed (52% and 65%, respectively).

Finally, although YEC's are more reluctant than other pastors to say "science should inform understanding of scripture, they strongly agree (84%) that *"The Christian community needs a greater commitment to showing how young earth creationism is consistent with science."*

#5: However, they are concerned about evolution for biblical reasons.

Over half of pastors said they had "major concerns" about the idea that God used evolution. The main reasons for that concern were that the idea "undermines the authority of Scripture" (64%), "views portions of the Bible as non-literal, like Genesis" (62%), "raises doubts about a historical Adam and Eve" (61%), and "raises questions about how and when death and sin entered the world" (59%). However, 26% of pastors saw no concern with the idea that God used evolution.

#6: The majority of clergy accept parts of scripture as symbolic.

60% of the pastors surveyed felt that "some portions of the Bible are symbolic, but all that it teaches is authoritative." Clergy whose views fall under theistic evolution and progressive creation were more likely to accept this statement (79% and 73% respectively), but a sizeable number of YEC pastors (40% among the core followers and 49% among those leaning towards YEC) also agreed with the statement.

#7: Clergy are concerned that changing their views on origins might compromise their ministry.

Over half of pastors (58%) who fell under the YEC category agreed that *"If you publicly admitted your own doubts about human origins, you feel you would have a lot to lose in your ministry."*

41% of pastors in the Progressive Creation group also agreed with the statement. Pastors who were uncertain or who fell under the Theistic Evolution group were less concerned, with only 26% and 17% respectively agreeing with the statement.

Dazed and Confused best sums up the findings of this research for me. Many of the clergy are uncertain and even those who believe the Bible will not commit to saying the Genesis narrative is fact, not an allegory. Many are closet evolutionists and are careful to avoid the subject of creation in sermons. The culprit is America's seminaries. As with the rest of academia, the influence in leadership has become largely secularized. Textual criticism is the 21st century toy that mesmerizes our Pastors and religious teachers and keeps them from absolute commitment to the scriptures. The absence of strong support for creationism from the pulpit has energized proponents like Ken Ham to take their message to the church.

This should cause grave concern for creationists and joy for evolutionists. Christian Pastors are being negatively influenced and are distancing themselves from a literal understanding the Genesis narrative. Many are not bold enough to declare their new-found faith in naturalism, so they avoid the topic all together. This silence is fueling the overwhelming success of teachers like Ken Ham. His success indicates that Bible believing Christians are hungry for believable, science-related information on creationism.

CHAPTER SEVEN

SEEKING ANSWERS FROM THE SOUL

W hose soul? Mine, of course. As a historian, teacher, and author, I must sort all of this out and inform my readers of the truth as I see it. Is the truth found in the previous six chapters? Is it buried beneath expert' bias and manipulated data? I believe that it is. Truth stands up for itself. It is a burning light that shines through the darkest night. Truth shouts from the housetops and mountains; it is not silent; it won't be silenced. Before I reveal my findings, let's look at the issues that influenced my reasoning.

#1 EVOLUTIONISTS HAVE A HISTORY OF MANIPULATION

It was the summer of 1926 in the sleepy little town of Dayton, TN. Here, giants on both sides of the creation-evolution issue collided in a cosmic battle for truth. Today we know the event as the "Scopes Monkey Trial". Creationism was represented by three-time democratic, populace presidential candidate, William Jennings Bryan. Evolution was represented by the famous agnostic criminal defense attorney, Clarence Darrow. H. L. Menken, the famous American author and cynical reporter for the Baltimore Sun, was present to heckle the Christians and cheer-on the defense.

The issue might surprise you, for in 1926 Tennessee it was against the law to teach evolution in the public schools. That's right, the Butler Act made it a crime! Now before we react to harshly, let's remember that it is exactly the reverse today, it is against the law to teach creationism in our public schools. As if by chance, a young substitute teacher, John Scopes, willfully violated Tennessee's Butler Act by teaching evolution. He was charged, arrested, and held in the town jail and an explosion ignited.

The thing most people do not know is that **it was all a set-up,** organized and financed by the American Civil Liberties Union as a test case in which Scopes had agreed to be tried for violating the hated Butler Act. The teacher of the class, a committed Christian, had refused to cooperate in the scheme. Scopes was charged on May 5, 1925, with teaching evolution from an unauthorized textbook entitled Civic Biology, which described the theory of evolution, along with its baggage, racial inferiority and eugenics. Scopes had predicted what would happen, encouraged students to testify against him and even coached them in their responses.

It was all a cleverly designed plan with an intent and purpose clearly in mind; one that brought about an 180 degree turn in the American public education system. From this point on, our courts, barraged with lawsuits by the likes of Madelyn Murray O'Hare, have gradually brought us to where we are today. Over the past 50 years, the ACLU has continued to keep IDC proponents from presenting their side of science from young minds; minds that should get all the information that is available on the issue, without prejudice or bias. Why are we afraid to allow students to make their own decision on Behe's 'mouse trap'? Is investigating that orbital time and atomic time may vary too much for college bound students? It is like evolutionists are so fearful that some bright mind will find the 'smoking gun' that, like Hitler and Stalin, they must suppress all contrary information.

In the end, Scopes was convicted of a misdemeanor and given a small fine, but the verdict was overturned on a technicality. The trial gained national attention and produced volumes of literature including a horribly biased drama, "Inherit the Wind", considered to be one of the greatest plays and later screenplays of the 20th century.

In the play, it seems that the court is overwhelmed by the scientific evidence supporting evolution, but in reality, the evidence presented in 1926 was mostly based in the theory of embryonic recapitulation, which has since been discredited.

The earliest argument proposed by the defense was that there was really no conflict between evolution and that the creation account in the Bible could be interpreted from the viewpoint of theistic evolution. In support of this claim, Darrow brought in eight experts on evolution, but other than Dr. Maynard Metcalf, a zoologist from Johns Hopkins University, the judge would not allow these experts to testify in person. Instead, they were allowed to submit written statements so that their evidence could be used in the case of an appeal.

Dr. Metcalf's testimony is enough for us to make several cogent points regarding these proceedings and the beginning of a conflict that would disenfranchise half of Americans living today who reject evolution. Dr. Metcalf, as would Richard Dawkins or any other evolutionist today, argued from the scientific information that was available to him at the time. Arguments that today have been challenged by both sides and are thought to be inaccurate, racially biased and unscientific. By the later stages of the trial, Clarence Darrow had largely abandoned the ACLU's original strategy and attacked the literal interpretation of the Genesis narrative and Bryan's limited knowledge of science.

Only when the case went to appeal did the defense return to the original claim that the prosecution was invalid because the law was essentially designed to benefit a particular religious group, which was thought to be unconstitutional and that is where we remain today. The trial revealed a growing chasm in American Christianity and identified two ways of finding truth, one biblically and one scientifically. It never occurred to people in that day that creationism also has science on its side. Liberals created a division between educated, tolerant Christians and narrow-minded, obscurantist Christians. The conventional view was that in the wake of the Scopes trial, a humiliated creationism retreated into the political and cultural background, a viewpoint evidenced in the play and movie *Inherit the Wind* and the majority of contemporary historical accounts. Objectivists saw the trial as a victory and not a defeat. Yet, Bryan's death soon thereafter created a legacy for his defense of scripture and left a leadership void that was eventually filled by other more qualified Christian apologists. Bryan, unlike others before him, brought name recognition, respectability, and the ability to forge a broad-based coalition of fundamentalist and mainline religious groups to argue for the anti-evolutionist position.

Nothing as contrived or blatantly coercive as the Scopes Monkey Trial should stand for truth. Those who support evolution and promote "Inherit the Wind" as great American literature should be ashamed of this fiasco. H. L. Menken, like the media of today, came to Dayton biased against creationism. His articles were scathing, unfounded and inaccurate. The truth is that this movie has been remade in two parts. The new titles are "Now You See Me" and the sequel "Now You Don't". So, the Scopes Trial itself offers us some insight in our quest for the truth. That objectivists are the ones that a closed-off, biased and afraid of the truth.

#2 EVOLUTIONISTS HAVE THE MIGHT AND THE MONEY

In today's America, any science supporting creationism is classified as religion, while any trivial theories, even alien seeding, that support evolution are celebrated as cutting-edge science. There is no fair court of arbitration; the federal government and the media are on their side. This should be true only if evolution is not a religion, which this author believes it is! This was demonstrated in chapter one; evolution is the religion of the objectivist or secular humanist. I know this is unsettling for those who believe that science is immutable truth and evolution owns science exclusively, but the fact is, evolution a theory, no matter what Collins, Dawkins or the Supreme Court says.

Evolution is too shallow to be protected as the only truth and the science that supports creationism should be allowed in our schools and public institutions. Today's students are bright and well-disciplined in their scientific pursuits. Why not let students have access to all the information and let them decide for themselves? Exposure to outside the box, radical thinking might just give them ideas for radical investigations that might change the world in which we live. Why are evolutionists afraid of competing ideas? Call me a conspiratorialist, but any theory that has the exclusive backing of the federal government, media and public schools and universities, is suspect in my book.

#3 EVOLUTION HAS BEEN USED TO SUPPORT SLAVERY AND SEGREGATION

Darwin's theory included the idea of racial inferiority and unleashed a tidal wave of racial abuse of Africans. Richard Weikart, author of a 2004 book, <u>From Darwin to Hitler</u>: <u>Evolutionary Ethics, Eugenics and Racism in Germany,</u> states:

> *Darwin clearly believed that the struggle for existence among humans would result in racial extermination. In <u>Descent of Man</u> he asserted, "At some future period, not very distant as measured by centuries, the civilised races of man will almost certainly exterminate and replace throughout the world the savage (African) races."*

Despite the good work that Darwin did in explaining "the survival of the fittest", his thinking on racial inferiority did much to encourage slavery and segregation by reinforcing racial stereotypes. Darwin proposed that blacks had lagged-behind whites in evolution. The problem is that this makes perfect sense if one believes in the evolution of modern humans from a variety of independent primate strains. These malevolent statements reveal Darwin's lack of a clear understanding of his own theory. Despite efforts by modern objectivists to dispel these inferiority notions, the root of Darwin's tree is despicably corrupt.

#4 EVOLUTION IS CONSTANTLY UNDERGOING CHANGE

Evolution has been through hundreds of modifications and revisions since first proposed by Darwin. Even today, many who claim to be objective scientists are embracing radical, complimentary theories, such as punctuated equilibrium and alien seeding, to make sense of its flawed science. Neither of these theories are based in science, by any stretch of the imagination. Punctuated equilibrium is the theory that answers creationists claims that transitional forms are largely absent in the fossil record. Alien seeding deals with the problem of not being able to create life in the laboratory, even with the best equipment and scientific minds in the world. This failure gives rise to the theory that human beings were placed on earth by aliens. This is science-fiction, at best. However, when the leading proponents of evolution suggest that either or both may be true, it is time for rational people to question its validity. Why don't our government and public institutions do the same?

Consensus does not establish truth! In the early 20th century the consensus in America, both north, west, and south, was that the Jim Crow laws (segregation of the races) were justified. The consensus was admittedly wrong and thankfully the wrong was eventually set right, but at what cost? The fact that there is a consensus among 'scientists' for evolution means nothing, especially when being a creationist in the academic and scientific community leads to not being tenured, being censured and losing one's position. Only scientists who are convinced that they are right would take such risks, but when they are silenced by our public universities and institutions, who is there to take their place? The academic and scientific communities have not genuinely considered contrasting views, especially when it comes to creationism. If you don't believe me, watch Ben Stein's film "Expelled". Once again, why are evolutionists so fearful of honest debate?

#5 DARWIN WENT TOO FAR

Darwin's discovery of macroevolution is brilliant and well supported in science, but in extending his theory across species lines without substanting evidence was reckless. That one idea is true does not make the other true, as well. For me, even after examining all of the information, the truth is obvious. Using "Occam's Razor", that no more assumptions should be made than are necessary, the idea of a supernatural creator, which requires but one assumption is more likely than evolution, which requires at least three absurd assumptions:

1. That with using the most sophisticated scientific equipment, laboratories and super-computers, scientists cannot produce life

in the laboratory, but assume that it happened by accident, in a pool of ooze, in a most inhospitable environment.

2. That in the absence of credible fossil evidence of transitional forms that clearly demonstrate species migration, scientists assume they exist and have even manufactured them or created beautiful drawings of them, which they sell as truth. Species do adapt and change demonstratively, but Darwin failed miserably in suggesting that dinosaurs became rodents and monkeys became men.

3. That the mathematical probabilities of evolutionary events falling perfectly into place, as evolutionists' claim, are completely unreasonable. The illustration of a bomb going off in a print type shop resulting in the production of a dictionary is not as absurd. Or how about a bomb going off in an airplane junkyard and producing a Boeing 747?

So, my reasoning is based in a systemic failure of the evolutionary community to accept responsibility for the discrepancies in their science and their unmitigated duality in defending those errors with subterfuge and deceit. They have refused to be professional or event honest in their public discourse, choosing to allow the US government and ACLU to fight their battles. Most have behaved the opposite of true scientists and scholars.

Understandably, objectivists have no choice but to believe in the pool of ooze, that somewhere genuine transitional forms might exist, and that the absurd mathematical improbabilities presented in DNA combination can be overcome; but they do not have the right to the moral high-ground and to attempt to silence all competing ideas. At least they could make the science add up for those of us who still have genuine, reasonable questions and concerns. Yet, this is a battle of public perception; media and governmental might over mind, so here is the judges score on the fight card:

Round #1 Public   Perception   Evolution

The media, US government and public schools have all imbibed the cool aid. The media has painted creationists as mindless, flat-earth,

Bible thumpers. They are despicable deplorables, clinging to their Bibles and religion. Those who send their children to government schools, allow those students to be shamed and intimidated into compliance. Creationist teachers and professors are silenced by the denial of tenure or losing their position. ID speakers are no longer welcome on college campuses. Creationists are relegated to a dangerous subculture that is bent on taking away the rights of women and the LGBT community.

Even though their scientists have not duplicated life in the laboratory, which would make their theory more palatable, they have, by sheer numbers, created a voluminous amount of narrative and illustration on the subject of evolution. One sees it in every newspaper, magazine, periodical, museum (except Ken Ham's), library, textbook, and even on the toilet doors. Their evidence, so called, is inescapable and overwhelming. But, does volume equate to truth? Did Hitler's "abundant, overwhelming evidence" that the Jew's were a "harmful infestation", make it true? Obviously, it did to the millions who followed him!

Round #2   Science   Tie

The sheer volume of the information produced by evolutionists and the fact that it is all that you hear or see in public media is overwhelming. How could so many eminent scientists and our public universities be wrong? Francis Collins, a fellow believer, is convinced of evolution from scientific evidence alone. Why would I question my brother in Christ? My visits to the Howard Hughes' Institute, near my home, have brought me into contact with sincere men and women, seemingly absent of a spirit of religious bias, who are proponents of evolution based on scientific evidence alone. Is it just their funding that they are protecting?

On the other hand, there is something to be said for the simplicity of Behe's 'mousetrap analogy'. The scientific explanation for the Earth not being nearly 13.7 billion years old deserve consideration. I've also been to the zoo and spent lots of time in the woods and although I find similar design elements humanity shares with the animals, I do not sense that the spirit of mankind could ever have entered a beast through some mystical process, yet unexplained. The fossil record is also on the side of the creationist. The science of creationism is profound, even though it has limited manpower and funding; it is convincing, even though the media debunks it out of hand.

Round 3   Philosophy   Creationism

Without the benefit of the Bible, Socrates, a nonreligious man, believed in a creator as did his famous and influential students, Plato and Aristotle. Even before becoming a Christian, Augustine of Hippo believed in a creator and it was not that he lacked a reasonable alternative, because the "Christian" Gnostics did not believe that God created the material world. They saw the entire creation as simply a divine hick-up, not an intentional act. Why didn't he join them?

Creationism is good philosophy because it produces beneficial actions. Creationists accept the uniqueness of mankind, which generates the noble expression of divinely initiated human empathy and compassion. Therefore, creationists build hospitals rather than euthanasia centers; orphanages rather than abortion clinics; food banks rather than mortgage banks; rehabilitation centers rather than welfare centers; and pregnancy advocate centers, rather than Planned Parenthood centers. The former Christian nations of Great Britain, France, Poland, and the United States opposed atheistic socialism in Germany, Italy, Spain, Japan, and Russia. Soldiers died for freedom and defending the weak. This kinder, gentler philosophy that focuses on helping humanity, is most appealing. The recent invention of the term 'speciesism' demonstrates how far the evolutionary community has gone to devalue human life. For me, saving human babies seems more philosophically relevant than saving baby seals.

Round 4   The Experts   Creationism

You can hate Ken Ham if you like, but he is doggedly immovable. One the other hand, Richard Dawkins folds like a cheap kite in a stiff breeze. Over and over again, when pressed hard on the vital question of the origin of life, he has evaded and quibbled. Behe has been challenged and maligned but he remains intractable. The creationist belief system is a product of something beyond knowledge, but shouldn't both sides be equally empowered by the truth they embrace?

Jewish, Muslim, and Christian creationists in China, Africa and even here in North America are being persecuted for their beliefs, but they remain committed and resolved. Years ago, when the Amish were being imprisoned for not putting their children in government schools, the Supreme Court decided that if individuals were willing to go to prison for what they believe, their choice is not a preference but a conviction of faith, and thus is protected under the First Amendment. When I see that there are academics who are willing to sacrifice tenure and careers for endorsing the science of creationism, I am drawn to that level of commitment. If things were reversed would evolutionists be as resolute? We will never know.

Round 5   Awareness   Creationism

One of the proofs for the existence of God is the Anthropomorphic Argument.

"To think at all is to think of God. One cannot think and not think of God, for God is inescapable. The Creator made a creature who is capable of thinking, and when the creature thinks he thinks of the Creator. We cannot do otherwise."

Written in the universe around man and written on the heart within man is the knowledge of God. This knowledge imparts a sense of accountability, not just a sense of fact. It creates an abiding awareness of personal responsibility to God, not just an intellectual knowledge of the being of God. For man to know of God is for man to know that he is answerable to Him. This is the essence of Paul's argument in Romans, Chapter 1 and it is widely recognized that throughout the world, even in the most remote parts. The creature is desperately seeking the creator. This is the primary reason for atheism; people do not want to be responsible or answerable to a Creator.

Written in the universe and on the heart of mankind is the knowledge of God. This imparts a sense of accountability, a personal responsibility to the Creator. This is the essence of Paul's argument in Romans, chapter 1, and it is widely recognized that the creature, throughout the world, is desperately seeking the Creator. The primary reason for atheism is rebellion that rejects any outside control or restrictions. In his book, The God Delusion, Richard Dawkins makes a feeble attempt to discredit what he calls 'the God hypothesis', but devotes only a few sketchy anecdotes to establishing that this God hypothesis is one that has defined religious belief through history and defines it around the world today. He fails to address the vital issues, as is usually the case with arguments he cannot win. I do not know where you are in your journey of exploring the wondrous miracle of creation, but if a still small voice is communicating that you were created with purpose, potential, design and significance, I suggest you listen.

My score card is 3-1-1 in favor of creationism. What is yours?

The truth is evolution won the fight in round 1 by a knock-out. The other rounds are superfluous! So long as evolutionary dogma is favored by the federal government and America's public schools and universities, with the expansive funding attached, creationists have about as much chance of winning this fight as did Little David with his slingshot against the mighty giant Goliath, and I guess that is the just the point. With such an advantage, why do half of American's surveyed still favor Creationism? They must hear that still small voice; the one that is the undeniable, anthropological proof that God exists.

EPILOGUE

By now you should have guessed that I favor the traditional creationist (OEC) viewpoint. Since I am a committed believer and am convinced that the scriptures are inerrant, I have no choice. Yet, I am always open and eager to listen to and consider all viewpoints. I enjoy studying and learning about alternative theories and have enjoyed sharing a few with you. Being closed minded is a waste and being closed off to new ideas keeps one from progressing towards the truth.

The purpose of this book was not to change minds, but to affect hearts. I wanted readers to be exposed to information that is not readily available elsewhere. For me, the jury is still out for the theories of Setterfield and Cremo, but deliberation and research must continue. Such complex topics, as the beginning of the universe, cannot be reduced to any single argument. Since one's viewpoint is ultimately determined by one's world view, it is difficult to remain objective. I get the most joy from non-believing IDC scientists who believe in creationism just because of the science.

To the objectivist, I pray for you to be spiritual enlightened, which is the only thing that can change your heart and mind about evolution. Might it be possible for you to imagine a beneficent Creator who loves you? Could you be less condescending to people of faith who do? Like you, they are a product of world view and are not mindless, evil people.

Look for some commonality as a bridge to compatibility. You might find that we are approachable and willing to exchange ideas without judgement.

To the young earth creationists, I implore you to consider that interpreting Yom as an era, rather than a day, does not destroy the veracity of scripture or damage the irrefutable arguments for redemption. If Peter's analogy of cosmic time is applicable to Genesis, the Gospel remains intact. Those who have sound scientific reasons for "millions of years" are not the enemy but are the only allies you have. Love them as Christ urged (John 13:35).

To my theistic evolutionary brothers and sisters, I say, "you can't have your cake and eat it too." The problem with your position has nothing to do with science but everything to do with soteriology. Your theory challenges the vital elements of redemption and unless cleaned up, leaves mankind hopelessly lost. Dr. Bonette has made a great start, but more scientific investigation needs to be made, if you would reconcile your viewpoint with Jesus' Gospel as revealed through Paul in the book of Romans.

To my traditional OEC companions I say, we must keep an open mind, explore all possibilities and be kind to those with whom we disagree. Young Earth Creationists are sincere, and their doctrine is born out of a determination to defend scripture. If Setterfield proves his theories, OECs and YECs might both be right. Treat objectivists with respect and remember you will never convert them with science, but through a chaste life filled with the Love of Christ. In 1980, I attended a lecture by a highly placed Director at NIH, who was a believer. He urged us not to try to convert agnostics by arguing the virtues of creationism but by showing them the love of Jesus. He pointed out that once a person comes to faith, he/she will get familiar with the Creator.

To those who still have an open mind and are not yet convinced of anything regarding creation, I encourage you to keep seeking. You may want to read the books and materials referenced herein. In the end, your decision on this vital issue will affect every aspect of your life. This is not simply a scientific debate but a battle for the soul. The vital cultural and social issues of today are all dependent on the determination of whether man is a cosmic accident and death is the end or that mankind was intentionally created with intent and purpose and that death is only a door to eternal life.

To all my reader's I wish for us to debate without malice or anger, and to seriously consider and respect each other's viewpoint. The great theologian and philosopher, Saul of Tarsus, wrote, "let us not look only on our own interests but also on the interest of others" (Philippians 2:4). This is great advice for those of us who want to know the truth about creation.

CPSIA information can be obtained
at www.ICGtesting.com
Printed in the USA
BVHW080135240221
600893BV00008B/712